D0614314

NOT FROM THE APES

NOT
FROM
THE
APES

Björn Kurtén

VINTAGE BOOKS

A DIVISION OF RANDOM HOUSE

NEW YORK

For Andrea

Copyright © 1972 by Björn Kurtén

All rights reserved under International and Pan-American Copyright Conventions. Published in the United States by Random House, Inc., New York, and simultaneously in Canada by Random House of Canada Limited, Toronto. Originally published in the United States by Pantheon Books, a division of Random House, Inc., in 1972. Originally published in Sweden as *Inte från aporna* by Albert Bonniers Förlag, Stockholm. Copyright © 1971 by Björn Kurtén.

Library of Congress Cataloging in Publication Data

Kurtén, Björn.
  Not from the apes.

  Translation of Inte fran aporna.
  1. Human evolution.   2. Hominidae.   I. Title.
[GN281.K813 1972b]   573.2   72-695
ISBN 0-394-71799-6

Manufactured in the United States of America
First Vintage Books Edition, August 1972

CONTENTS

PROPOSITIONS

I did not set out to "prove" the following theses. On the contrary, they were formulated gradually in the course of my work, and many of them are contrary to my own previous beliefs.

1. The ancestry of man on one hand, and the apes and monkeys on the other, has been separate for more than 35 million years.

2. Man did not descend from the apes. It would be more correct to say that apes and monkeys descended from early ancestors of man. The distinction is real: in the traits under consideration, man is primitive, apes and monkeys are specialized.

3. Our ancestors were tree-living up to about 10 million years ago; they descended from the trees between 5 and 10 million years ago.

4. Our ancestors were not forced to leave the trees by some kind of crisis, such as death of the forest by desiccation. They came down to the ground to invade a new, favorable life zone.

5. Differences in sexual behavior may have been over-stressed. There is reason to suspect that there was no absolute division between foraging-hunting-fighting males, on the one hand, and stay-at-home, baby-sitting females, on the other.

6. Mankind has repeatedly split up into distinct species of which all but one died out, presumably by competition if not by direct struggle or fighting with one another.

7. Natural selection is still a vital force in human evolution. Far from having disappeared, selection by differential mortality still exists, and its importance will probably increase.

The typescript was read by Drs. W. W. Howells and S. J. Gould (Cambridge, Massachusetts) and C. S. Coon (Gloucester, Massachusetts). I gratefully acknowledge their comments and editorial help. The remaining errors are my own.

Björn Kurtén

Cambridge, Massachusetts
December 1970

PART 1

THE RECORD

MAN IN
THE PETRIFIED ZOO

Probably very few people are aware of the magnitude of geological time or the fabulous richness of its preserved record of the life of the past. The millions of years and the millions of species that are dead and gone defy comprehension. Perhaps a visit to one of the great geological museums will give the spectator little more than a confused impression of a multitude of bizarre bony faces; there are indeed shapes and expressions galore, from the rapacious grin of the tyrannosaur to the lugubrious, long face of the duckbill dinosaur in its unbelievable top hat.

But there may also be more familiar faces in this petrified zoo: faces human and semi-human. We may take note of the strong, clean lines of the skull of Cro-Magnon man; of the owlish stare of the Neanderthaler's head with its large, round eye sockets; of the scraps and fragments of teeth, jaws, and braincases of still older half-men and pre-men of the early ice age back to one or two million years ago.

Such finds are rare, *for men are generally too smart to become fossils.* It is, so to speak, the bungler who risks becoming a fossil. He may get caught in quicksand or fall down a pothole; he may come too close to the undercut bank of a river swollen in flood; he may get his head taken off by a rival tribe, his brain eaten and his skull thrown in the refuse heap. This is how men, and the ancestors of men, became fossils—up to the time when deliberate preservation by burial became a custom.

Thus the remains of man-like beings are few, and in general they are also late in geological time. The Cro-Magnons lived some 20,000–25,000 years ago; the last Neanderthaler died probably some 35,000 years ago. Dates for the early human and semi-human types may run to a million years or two; in fact, some of these are now dated at more than five million years. This is awesome and yet little more than a scratch on the surface. The tyrannosaur and the top-hatted dinosaur are among the last of the dinosaurs, yet they lived more than 65 million years ago. And the age of the earth is currently estimated at nearly five billion years.

No human being ever saw a living dinosaur, but our history did not begin five million years ago. Until quite recently it could still be said that man arose from something pretty close to the ancestry of the gorilla just before the beginning of the ice age (3–4 million years ago). In that case the early men should be more and more ape-like as we go back in time. Surprisingly, they are not; indeed, the term "ape-men" is quite misleading for these early humans. They point back to a very different sort of ancestor.

In fact, it has been possible in the last decade to demonstrate that the human lineage can be followed back into far more distant times where it still retains its unique human character. The fossil evidence now makes it most likely that the family of man has been distinct for at least 35 million years—more than half the time elapsed since the last dinosaur died. Indeed, we may doubt that our ancestor was ever what could properly be called an ape. This makes excellent sense zoologically. The contrasts between apes and men in anatomy (for instance of the teeth) are too great to be reconciled with a relatively recent common origin, and the same is true of behavior.

It is my aim in this book to review the evidence as I see it, and to present it in a non-technical form accessible to every reader. Of course, the conclusion that man arose from a lower form of animal life has lost some of its power to agitate people since the heroic days of early Darwinism, and we might perhaps be tempted to wonder why it should matter whether we come from apes or not. I believe it matters for the simple reason that knowing whence we came will help us understand where we are going.

NAMES

Many books about fossil men present the reader with a staggering number of "scientific" names to keep track of, suggesting that practically every one of our ancestors was a species of his own. Early finds of fossil men were indeed so rare that the proud discoverer was prone to emphasize their distinctness and importance by giving them a new species or genus name. Unfortunately, the tradition has continued and is still practiced even now although the number of known fossils has grown rapidly. For instance, recently an almost indeterminable skull fragment was dubbed a new genus and species although the fossil probably belonged to a species already known and there was no morphological detail that could be used to distinguish it. Zoological systematists have by now become quite wary of encumbering the formal nomenclature with new names on insufficient grounds. It is hoped that a new generation of anthropologists will see the practice in the same light.

In this book formal names will be kept to the minimum that seems to me to correspond to the actual biological diversity revealed by the fossil finds. In this, I am steering a middle course between the so-called splitters and lumpers. The splitters tend to see evidence of a large number of species and genera. Lumpers, on the other hand, regard much variation as simply individual, with no systematic significance. I shall have more to say on this topic later on, when we come to the interesting problem of whether mankind ever formed more than one contemporary species.

However, to help readers who may be familiar with some of the names not accepted here (they may be referred to as synonyms), I have collected them in the glossary at the end of the book, with some remarks on what they stand for.

At this point we have to introduce some basic facts about systematics that are necessary for the understanding of the text to follow. The species is the basic unit in classification. A species at any given time is formed by a population or group of populations that are interbreeding, or at least have the potential to do so. A good example is living man, the species *Homo sapiens*.

A species name consists of two parts, of which the first is the name of the genus to which the species belongs (*Homo* in our case), and the second is the trivial name of the species (in this case, *sapiens*). The use of such formal names is governed by a set of internationally accepted rules. The primary aim of the rules is to ensure stability in nomenclature, which seems most praiseworthy at a time when many discussions fall into confusion because contending sides use the same words in different senses or vice versa.

The original "meaning" of the names, on the other hand, does not matter. The first describer of the species may have been in error about the true character of the form of life that he was describing and so may have coined an entirely inappropriate name. There are some who would impute this error to Linnaeus, who established the name *Homo sapiens* (Man, the wise). Whatever its meaning, the name given by the first describer is regarded as valid.

Groups of related species are put in the same genus: for instance, *Homo sapiens* and the extinct species *Homo erectus* (Java man, Peking man, etc.) are now generally regarded as so close together that they belong in a single genus. Somewhat more distantly related forms are put in other genera but still belong to the same family, the Hominidae—men and pre-humans.

Next to the Hominidae, or hominids, in the scheme of nature stands another family, the Pongidae, which accommodates the living apes (gorilla, chimpanzee, orangutan, and gibbons) and their fossil allies. The Hominidae and Pongidae together with several other families (the monkeys, lemurs, spectral tarsiers, and various extinct groups) form the great order Primates, which is one of the main divisions of the class Mammalia, or mammals.

The hierarchy is given greater flexibility by the use of intermediate categories such as super- or subfamilies. The family Pongidae, for instance, comprises two rather well-differentiated groups—the great apes on the one hand and the gibbons on the other—and these are put in separate subfamilies (Ponginae and Hylobatinae respectively).

A species may also be divided into subspecies, which are the smallest systematic units to receive formal names. The name of a subspecies is made up of three parts—the first two are the same as that of the species to which it belongs, and the third is its own subspecific trivial name. If, for instance, you think that Neanderthal man was a distinct subspecies but belonged to the same species as we, you call him *Homo sapiens neanderthalensis* (in which case all living men might be regarded as another subspecies that must then be called *Homo sapiens sapiens*). But subspecific division of *Homo sapiens* is tricky, for various reasons, and this category will not be used much in this book.

What should be kept in mind is that the names we use are intended to reflect our ideas about the biological relationships among the creatures we are talking about, be they men or apes, and that is the real significance of formal nomenclature.

FINDING
THE DARTIANS

The main types of extinct men known today are the Neanderthal man discovered in 1857, Java man discovered in 1892, and the Dartian, or *Australopithecus*, discovered in 1924. Of the three, the Dartian is the oldest and most primitive. Thus, the Dartian must be the pivot of our discussion; his physical characteristics are the most likely to throw light on the ancestry of the hominids. Furthermore, we must try to form some ideas regarding his probable habits and mode of life, so as to help us towards an understanding of the adaptations and life of his predecessors.

The first fossil of a Dartian to come under scientific study was the skull of a child from a limeworks at Taung, (Ta-ung, place of the lion) in what was then Bechuanaland and is now known as Botswana. It was revealed by blasting during excavations and sent to Professor Raymond A. Dart in Johannesburg. The skull turned out to be very well preserved: there was all of the face, the lower jaw, and the right side of the braincase as well as a natural cast of the brain cavity. The teeth were milk teeth, but the first permanent molar had also been cut. This happens in man at an age of about six years, which would thus be the approximate age of the Taung child. But the creature was not a man, nor was it an ape; it was a very primitive being with a brain much smaller than that of any human child of comparable age.

Publishing the find in 1925, Dart named it *Australopithecus africanus* (the southern ape of Africa)—which he subsequently regretted. As he points out, a name like *Homunculus* would have been more suitable, but the rules of zoological nomenclature do not permit such a change (and incidentally, the name *Homunculus* had already been used for a fossil monkey from South America). In 1947 Sir Arthur Keith suggested the colloquial name "Dartians" for *Australopithecus africanus* and its relatives. I have followed his lead here. The usual colloquial term, *Australopithecine,* is not only clumsy but also misleading, as it suggests that these early men form a distinct subfamily of their own.

Sir Arthur thus paid homage to Dart, who "not only discovered them [the Dartians] but also so rightly perceived their true nature." The situation was different in the twenties. Anthropologists then were not, as a rule, ready to accept Dart's conclusions regarding the importance and probable role in human evolution of *Australopithecus*. But later discoveries and studies have quite vindicated them on all essential points, and there is little reason for going further into that old controversy. As we shall see later on, practically all of the epoch-making discoveries of new kinds of fossil hominids have been received with doubts and opposition from most contemporary anthropologists.

Robert Broom, a doctor and paleontologist who had spent a lifetime studying the fossil reptiles of the South African Karroo—a much earlier stage in man's evolution —was not satisfied with the way in which "Dart's baby" was explained away (it was commonly regarded as something akin to the gorilla and chimpanzee), and decided that the time had come to find additional specimens. The fossils from Taung had come out of an ancient cave filling, and interest had originally been aroused by the find of fossil baboon skulls. Now fossil baboons were reported from the Sterkfontein Valley in the Transvaal. Broom followed the trail.

At seventy he set out to search for fossil pre-humans in the Sterkfontein Valley. When he died at eighty-four in 1951, he had succeeded beyond his wildest hope. Remains of at least fifty individuals were known then, and the total is now over a hundred. This means that we now are quite well informed about some of the anatomical traits of the Dartians—especially the teeth, jaws, and braincases—and it is now fully clear that they are hominids and have nothing to do with the apes.

Broom started work in 1936 and found a good specimen within three months: a skull of an adult Dartian from an ancient cave, at the Sterkfontein Type Site. This locality has since yielded many more finds. The kind of Dartian found here appears to be similar to that from Taung, although at first it was given a different name.

Later, several other finds of pre-humans were unearthed from two other ancient caves in the Sterkfontein Valley. These sites are called Kromdraai and Swartkrans. The specimens from these localities differed from the Taung-Sterkfontein type in many respects and may be known under the name *Australopithecus robustus*. As the name implies, these creatures were more robustly built than the Dartians from Taung and Sterkfontein, and many students now regard them as a separate species. I accept this conclusion for the moment and will return to the species problem in a later chapter. The two forms will be referred to as the gracile and robust Dartians respectively.

Gracile Dartians were again found by Dart at a site (the Limeworks cave) near Makapansgat in the central part of the Transvaal. Like the previous finds, these came out of an ancient cave filling, which meant either that the Dartians had died in the caves or that their remains had been dragged in by someone else.

Dating the Dartians proved difficult. The only thing that could be definitely shown was that the sites with gracile Dartians were older than those with robust Dartians. It now seems probable that some of the latter may be as young as half a million years while the older sites with gracile Dartians may be at least one and perhaps two million years old. This led to the assumption that the gracile Dartians evolved into robust Dartians and that the whole lineage was a side branch in human evolution.

The location of the finds might also suggest such an interpretation. As long as Dartians were found only in southern Africa, they might seem a side branch that evolved in this out-of-the-way area far from the mainstream of human evolution. But with the recognition of Dartian fossils from other parts of Africa and possibly from Asia as well, it became clear that the robust and gracile forms had indeed coexisted for a very long time and that the Dartians did represent a stage in the evolution of man. At the same time, definite dating became possible.

The crucial evidence came from Olduvai Gorge, a canyon in northern Tanzania which cuts through a sequence of strata spanning the greater part of the Pleistocene epoch (the last two or three million years in the earth's history, including the ice age). Thanks to the indefatigable work of L. S. B. Leakey, a remarkable harvest of fossils of men and beasts and of man-made tools have been discovered there.

Lava and tuff deposits in this sequence have been dated by the potassium-argon method, the most important of the so-called radiometric dating methods at the present time. It is based on the fact that natural potassium (a very common element in nature) contains a fraction of a radioactive isotope, which when decaying produces certain amounts of the gas called argon. This goes on at a fixed time-rate, and hence the contents of potassium and argon in a given rock yields a measure of the time since the rock was formed. The method is reliable in the sense that it gives consistent results, and as far as the earlier dates are concerned—from, say, ten million years on—the potassium-argon dates are corroborated by other methods. There remains some doubt regarding the later dates, especially those of less than a million years. For very latest dates there may be great discrepancies, e.g. a potassium-argon date of 240,000 years for cultural material that by all other dating methods cannot be older than about 30,000 years.

Dartian fossils, both of the gracile and robust type, occur in some of the oldest deposits found in Olduvai Gorge, dated at some 1.7–1.8 million years. The Dartians range up into much younger strata; according to a somewhat uncertain identification, robust Dartians were still in existence in this area less than half a million years ago (this would be approximately contemporary with the robust Dartians in the Sterkfontein Valley). The gracile type, however, vanished at a much earlier stage, probably because it was gradually transformed into a higher form of man (*Homo erectus*). In strata with an age of about one million years, we find the remains of transitional forms that may be advanced Dartians or primitive *Homo erectus*. This gives an idea of the probable evolutionary role of the gracile and robust Dartians respectively: the one ancestral to higher hominids, the other a barren side branch.

There are various other finds of Dartians from Africa, but the most important ones come from Omo Valley at the northern end of Lake Rudolf in Ethiopia. Here is found a sequence of strata going back to about four million years, and Dartians have been found at several levels from about two million years backwards in time to almost four million. Dating by the potassium-argon method is based upon volcanic tuffs intercalated at various levels in the sequence.

Preliminary notes on the fossils have recently been published by F. C. Howell, who shows that both the robust and the gracile type are represented. The period of coexistence of the two forms of Dartians is very long indeed, two to three million years at least.

At the time of this writing, the oldest known fossil of a Dartian appears to be a jaw fragment of the gracile form found near Lothagam Hill at the southern end of Lake Rudolf. Bracketing potassium-argon dates from overlying and underlying tuffs indicates an age between four and eight million years. The rich fauna of fossil mammals found in association with the Dartian has a distinctly "earlier" stamp than known East African fauna dating back to four million years, and the Lothagam jaw is thought to be between five and six million years old.

Were Dartians present outside Africa, too? The evidence is not yet accepted by all the authorities. There were finds from the Levant (Ubeidiya) and from East Asia (fossil teeth bought in Chinese drugstores [see p. 32], of unknown provenance) which are regarded by some students as Dartian. Perhaps the most interesting fossils are the finds from Java that have been called *Meganthropus;* a few jaws and some isolated teeth have been found, and they certainly resemble robust Dartians. The age would be about right, too, for they may be more than a million years old, perhaps upwards of two million.

THE TEETH
OF SUPERMAN

Dartians have many traits that at first sight seem ape-like—such as the small brain, the eyebrow ridges, and the protruding jaws—but if we look closer at them we must conclude that they are simply primitive, as compared with ourselves, rather than specifically apish. The brain, for instance, is much smaller, on an average, than that of modern man. The average capacity of the Dartian braincase is about 450 c.c., or one-third of the average modern man (about 1,350 c.c.). The Dartian figure is not much different from that of the gorilla, but of course this does not make the Dartian ape-like: he is just primitive. Due to the small size of the brain and the large size of the jaws, the Dartian skull profile is not unlike that of an ape; however, if we look at the details, we find that almost all are different. This is especially clear when we look at the teeth.

The ape has large, powerful canine teeth (eyeteeth) that jut a long way out over the other teeth, making it necessary to have empty spaces in the opposing tooth row for the reception of the tusks. There is such a space in the upper jaw between the canine and the front teeth, and in the lower between the canine and the first cheek tooth. The canine itself, if you imagine it sawn off at the base, has a cross section that is elongate from front to back.

In the Dartian, on the other hand, the canine is small and does not jut out over the other teeth, so there is no need for empty spaces: the tooth row is closed, as in modern man. The shape of the canine is much like ours —slightly pointed when quite unworn, then more or less spatulate like the front teeth. Its cross section at the base is also human, and is wider (transversely) than long in contrast with the elongate shape of the ape tusk.

Look next at the front teeth, or incisors. In the apes these are very big, and the front part of the jaws, consequently, very broad and muzzle-like. Not so in the Dartian, whose incisors are if anything even smaller than ours.

If we derive men from apes, then the Dartian has superhuman canines and incisors: the teeth of Superman in the head of the pre-human. But of course he does not really: he is just pointing back to an ancestral type quite different from the ape.

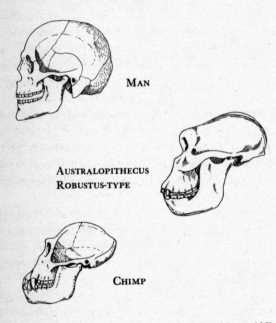

MAN

AUSTRALOPITHECUS
ROBUSTUS-TYPE

CHIMP

FIG. 1   Skulls of modern man (top), robust Dartians (middle), and chimpanzee (bottom).

We come now to the cheek teeth. Men and apes have the same basic pattern in their molars, a heritage from a common ancestor long ago, and there has been comparatively little modification of it. But when we look at the two foremost cheek teeth—the premolars, just behind the canines—remarkable differences are seen, especially in the lower jaw.

In the apes, the premolar just behind the lower canine tusk is very oddly shaped. It is much bigger than the premolar next to it and markedly elongated from the front to the back. It normally has a single, pointed cusp although there may be a small internal cusp in addition. A sharp keel runs from the tip of the main cusp down the front edge of the tooth. It is evident that, together with the upper and lower canines, this premolar forms a functional complex by creating an interlocking cutting apparatus. The premolar behind it has an entirely different shape. It is moderate in size, relatively short and broad, and two-cusped—with an external and an internal cusp.

In man and in the Dartians, both the lower premolars are of this latter type. They are more or less rounded in cross section, not elongate, and have two cusps—external and internal.

The tooth row as a whole, in man and in the Dartians, has an arcuate or parabolic shape; the cheek teeth diverge from each other towards the back of the jaws. In the apes, on the other hand, the tooth row is U-shaped with parallel cheek teeth. The reason is that the front teeth are much larger in the apes and the muzzle broader in front than it is in the hominids. To accommodate this great width of the jaws in front, most apes have developed a special strengthening bony structure—the simian shelf—connecting the left and right branches of the lower jaw. This shelf is lacking in man and in the Dartians. The shape of the palate is strongly arched in man and in Dartians but flat in the apes. The smallness of the incisors and canines in the Dartians has already been noted. The cheek teeth, however, are very large.

Combining all these facts, we can make some suggestions about the probable kind of dentition in the ancestors of the Dartians. It seems likely that they had small incisors and canines, two-cusped lower front pre-molars, and relatively large cheek teeth. The tooth rows probably diverged backwards and were closed, without marked empty spaces between teeth.

The interlocking canine teeth of the ape also restrict the movements of the jaw in chewing, and as a result the wear of the cheek teeth is quite different in apes and men. While the wearing surface of the ape teeth becomes more or less irregular, that in man *is perfectly flat* as a result of the more complex jaw movements. Again, the same is found in the Dartian.

In the robust Dartians the molars are very big, and the size of the muscles that operated the jaws must have been huge, for a bony crest is formed on top of the skull to give a sufficient area of attachment. This has been cited as an apish characteristic in the Dartians. On closer study, however, the way in which this crest is formed underlines the difference from the apes.

It is the big temporal muscle, one of the main muscles that close the jaw, that is responsible for the appearance of a crest on top of the head. In the apes, with their prominent front teeth, the muscle passes obliquely backwards and upwards from the ascending branch of the lower jaw to the occiput, or upper hind corner of the skull. Here the fore-and-aft, or sagittal, crest is formed for their attachment, and it is confluent with the great transverse crests bordering the hind edge of the skull. These are called the nuchal crests and serve as attachment areas for the neck muscles that anchor the head to the body.

In man, the main direction of the temporal muscle

is vertical rather than oblique, and its function is mainly concerned with the molar part of the dentition rather than the front part. If we let this muscle grow to the size it had in the robust Dartians, it will extend upwards towards the crown of the head rather than the occiput. Finally the two muscles of the left and right side will meet on top of the head and a crest will form here for their attachment. This crest is now quite distant from the nuchal crests; furthermore, the latter have shifted backwards and downwards to a great extent in relation to the upright posture of the Dartians. The position and development of the sagittal crest in the Dartians must then be regarded as a uniquely hominid characteristic. Once again we are faced with the suggestion of a particular type of dentition in the ancestors of man: a type with small, non-interlocking canines, small incisors, large molars, and a correlated musculature which makes for powerful chewing but weak biting powers.

STANDING UP

No ape is a biped in the sense that man is. The apes are adapted to life in the trees, swinging by the arms and moving hand over hand. Because of this, their arms have become lengthened but their legs shortened; as a result, when walking on all fours, they tend to have a semi-erect posture. Some apes, notably the gibbons, can run erect on two legs balancing themselves with their long arms held out, but this is obviously something quite different from the two-legged gait of man.

In the case of the Dartians, the evidence for a man-like, erect posture is quite compelling. This is shown in the skull, for instance, by the position of the neck opening, or foramen magnum, through which the brain connects with the spinal cord. The opening is flanked on both sides by the two condyles forming the joint between the head and the neck. In modern man this opening is centrally placed on the under side of the skull so that the head balances almost perfectly on top of the vertical neck, thus saving unnecessary tension of the neck muscles. Actually, as most of us know from personal experience, the balance is not quite perfect; if we go to sleep in a sitting position, the head will drop forwards as the neck muscles relax. In other words, the center of gravity of the head is still somewhat in front of the condyles of the neck joint. However, the necessary muscle force is slight and no long leverage is needed; accordingly, the nuchal attachment area for the muscles is situated quite far down, near the lower hind corner of the skull.

In the apes, on the other hand, the neck opening faces obliquely backwards, showing that the head did not balance on a vertical neck, and the attachment area for the great neck muscles needed for the anchoring of the heavy-jawed, protruding head is high up on the skull.

In the Dartians the position of the opening is very close to that in modern man, and it faces straight downwards, showing that the neck was vertical. The head, however, was not as well balanced as in modern man, and relatively powerful neck muscles were present, as shown by the wider leverage and larger size of the nuchal attachment areas. In this respect, then, the Dartians appear to have been somewhat closer to our four-legged ancestors, and we may get the impression of an unfinished adaptation.

But although the inheritance from ancestral forms is

more evident in the Dartians than in ourselves, this does not mean that the adaptation as such is "unfinished." In fact, it would be impossible for a creature to achieve a perfect balance of the head if he needed jaws of the size found in the Dartians and had a brain as small as theirs. To achieve balance, the foramen magnum would have to shift much farther forwards, in relation to the braincase, than in ourselves. This would probably lead to serious difficulties for skull and brain architecture. In any case, if you are big-jawed and small-brained enough, you have to rely on the strong muscles at the back of the neck.

The next part of the skeleton in which erect posture is clearly reflected is the hipbone. In the apes it is long and narrow; in man short and broad. The iliac bone, which forms the large "blade" of the hipbone, is short and broad in man, and its outer parts are turned forwards; in the apes it is straight, long, and narrow. The difference in shape is related to the curvature of the human backbone. It curves backwards in a sway-backed fashion just over the hips, and then curves forwards higher up like an extended *S*. In contrast, the ape spine curves forwards only and there is no trace of the sway-back.

The broad blades of the iliac bones in man serve as a support for the intestines and as areas of attachment for the powerful gluteal muscles that shape the human buttock—a typical characteristic of man. With man's erect position, these muscles have taken over the task of balancing the trunk on the legs. As a result, the hamstring muscles, which originate beneath the hipjoint, lose some of their importance in man; in fact, the lower (or ischial) part of the hip girdle is relatively smaller in man than in the apes.

The hipbone in the Dartians is basically similar to

that in man, with a broad iliac blade and a reduced ischial portion. Differences from modern man are slight but interesting, for they may be correlated with differences in gait. The iliac blades, for instance, are splayed apart in front to a greater extent than in man—this is a primitive characteristic. The resulting "openness" of the belly is compensated for, to some extent, by a special projection from the front of the iliac bones; this has been lost in modern man. The whole arrangement suggests to some students that the Dartian spine was somewhat less sway-backed than our own. It has also been thought that the Dartians were not capable of a truly human striding gait. The gracile Dartians, however, would seem to have been fleet runners. It may be questioned whether the more shuffling gait was any serious drawback to gracile Dartians, for they were small (less than four feet tall) and correspondingly light-bodied.

Remains of the thighbone are incomplete but do show that the Dartians had already acquired the slightly knock-kneed stance typical of man: the thighbone slopes inwards from the hip while the shin is vertical.

The foot of a Dartian found in early strata at Olduvai shows a basically human structure quite unlike that of the hand-like foot of the apes. The big toe, for instance, was not opposable to the other toes as in apes (whose feet functionally resemble our hands more than their hands do). Yet the Dartian foot appears to be more primitive than ours, for the big toe is not as predominant in size. The second and especially the third or middle toes are well developed and they probably took part of the weight in walking, more so than in our foot, in which the big toe does the main work.

While the ankle joint of the Dartians was typically human, the knee joint may have been somewhat more primitive as suggested by the position of the muscle scars found on the shin bone; however, this arrangement does not specifically point back towards an ape-like condition.

Only a few remains of the hand have been found. Those of gracile Dartians indicate a light build and a much more flexible hand than in apes, although not quite as flexible as in modern man. The hand of robust Dartians was apparently a great deal more powerful. The Dartian thumb was evidently quite opposable as in ourselves, and would serve well for the "power grip" of the whole fist and reasonably well for the "precision grip" of the thumb and index finger. This latter grip is impossible for the apes, who have reduced thumbs and greatly lengthened fingers to facilitate rapid hooking of the hand over branches when moving hand over hand.

If we try to gather these facts into a comprehensive picture, we may now see the Dartians, especially the gracile form, as agile, fully erect creatures, perhaps slightly less sway-backed than modern man. They also seem to have been excellent at running, perhaps slightly clumsier at walking (although with their small size and lightness this effect might well have been negligible), and had nearly or fully human hands for the manipulation of tools, weapons, and other objects. Some anatomical details may seem less "finished," or finished in a slightly different way, when compared with those in ourselves. It would seem clear that the Dartian exhibits more of the four-footed ancestry than we do, and it is of course certain that he evolved from four-footed ancestors. How and when this transition occurred we do not know. In any case there is no reason to assume that there was a transitional stage of the stooping type peculiar to the apes (a special character of the pongids). Their hand-like feet are not set flat on the ground but rest on the outer side in a peculiar bowlegged fashion completely different from the knock-kneed stance of man. It appears most unlikely that any human ancestor ever walked in such a fashion. The same certainly holds true for knuckle-walking, which is unique to the living African great apes —the chimpanzees and gorillas. It has been suggested that this type of gait developed from an original fist-walking habit. There is nothing to suggest that any ancestral hominid had this habit.

RAMA'S APE

The earliest Dartians we know about lived some four million years ago. This carries us well back into the Pliocene epoch, the last subdivision of the Tertiary period. Ten years ago very few students would have accepted the notion of Tertiary hominids. And yet the evidence was already in hand, for although the 4-million-year old Dartians were not then known, *Ramapithecus* (named for the Hindu deity Rama, symbol of exalted purity) had been described as early as 1934.

This genus, named by G. E. Lewis, was used to describe the upper jaw of a small primate found by the Yale Expedition to the Siwalik Hills in northern India. This range of hills contains thick fossiliferous strata which have long been known and studied by geologists. They were laid down millions of years ago by a predecessor of the Indus and Ganges rivers. It has been thought that this was a single river that, in ancient times, flowed out from the rising Himalayan range in the region of present-day Assam. From there it flowed northwest along the trough formed at the foot of the mountain range, and finally turned south towards the Indian Ocean, following the same course as the present-day Indus.

The river probably came into being with the first great folding and uplift of the Himalayas during the Miocene epoch, some 20 million years ago, and continued throughout the Pliocene epoch beginning about 12 million years ago. Finally, during the Pleistocene epoch, the Siwalik Hills area was elevated to its present height. As a result, the ancient Siwalik River was forced to change its course and become the Ganges, and the connection with the Indus was lost.

The great series of strata in the Siwaliks thus begins with the later Miocene and continues throughout the Pliocene and well into the Pleistocene. Enormous numbers of fossil remains have been excavated at all levels. They are the bones of animals that died by the riverside or perhaps drowned in the river. Many of the bones have been mutilated by crocodiles, others show evidence of gnawing by hyenas.

The jaw of *Ramapithecus* was found at a level about halfway up the sequence, in the so-called Nagri formation. The same appears to be true for most subsequent finds of this creature from the Siwaliks. The age of *Ramapithecus* may thus be set somewhere in the early Pliocene, probably in the range of 8–12 million years. Some specimens, however, may date back to the later Miocene (12–14 million years).

Lewis noted some remarkably man-like traits in his new fossil, traits definitely lacking in the numerous ape jaws of the same age known from the Siwaliks and also from Europe and Africa. He was bold enough to classify *Ramapithecus* as a member of the human family, the Hominidae. It was the first description of a Tertiary

A. RAMAPITHECUS

B. ROBUST DARTIANS

C. GRACILE DARTIANS

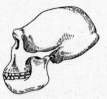

D. HOMO ERECTUS

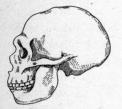

E. NEANDERTHAL MAN

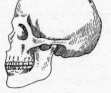

F. HOMO SAPIENS, including
modern man

FIG. 2 If we include *Ramapithecus* (A, restoration of
skull) , we now know of six main types within the family
Hominidae. The five others are B, robust Dartians; C,
gracile Dartians; D, *Homo erectus;* E, Neanderthal man;
F, modern man.

hominid, and it should have been the sensation of the
day.

Shortly afterwards, Lewis' conclusions were corroborated by the noted authorities on primate denti-

tions, W. K. Gregory and M. Hellman. A joint study of the primates from the Siwaliks was published by Gregory, Hellman, and Lewis, and here the resemblance between *Ramapithecus* and man was again stressed.

In spite of this, hardly anybody took notice. Attention was directed by the remarkable finds of Peking and Java man, and soon afterwards by Broom's new Dartians from the Sterkfontein Valley; in the controversy that raged over these more recent finds, unassuming *Ramapithecus* was overlooked.

The only anthropologist who bothered to examine the specimen was Ales Hrdlicka, a leading figure in American anthropology. His view of presumptive fossil hominids may have been colored by his experiences in America, where numerous indigenous remains of "ancient men" had been submitted to his scrutiny. He invariably unmasked them as quite recent. His approach to *Ramapithecus* had some of the same quality of unmasking. This form, of course, was undeniably fossil although Hrdlicka refused to accept it as a hominid. He regarded it as an ape with some manlike traits.

In fairness to Hrdlicka, it must be remembered that most anthropologists at this time regarded the Dartians, too, as apes. Actually, Hrdlicka, when referring to *Australopithecus*, stated that *Ramapithecus* was the more manlike of the two!

In the thirties scientific thought was not ripe for such a staggering notion as a Tertiary hominid. There were enough problems to sort out in the Pleistocene, and so *Ramapithecus* was forgotten. In the late fifties, however, E. L. Simons resumed study of this form. He was soon able to identify other, hitherto unrecognized, specimens of *Ramapithecus* in various collections from the Siwaliks. He also showed that a rather ape-like lower jaw that had been assigned to *Ramapithecus* by Gregory, Hellman, and Lewis in fact had belonged to an ape, but that some other jaws with man-like traits clearly belonged to *Ramapithecus*.

Most of the specimens have since been shown to come from the Nagri formation and so are approximately of the same age. Simons showed convincingly that Lewis' claim of hominid relationships for *Ramapithecus* was well founded, and this extinct primate is now widely acclaimed as a member of our own zoological family and as the probable ancestor of the Dartians.

Meanwhile, another upper jaw of *Ramapithecus* was found by Leakey at a site called Fort Ternan in Kenya. It is closely similar to the Indian form, but evidently is quite a bit older, for it has a potassium-argon date of 14 million years.

But *Ramapithecus* appears to have ranged even more widely. A number of fossil teeth from a coalfield in the district of Keiyuan, Yunnan Province, China, show the characteristic features of *Ramapithecus* and evidently represent an eastern tribe of this form. There are also finds from Europe that appear to belong to *Ramapithecus*. One of them is a tooth from southern Germany; other specimens, from Hungary and Greece, still await description. Both the Chinese and the European specimens evidently date from the early Pliocene, so that they are of the same age as the Indian ones. The African finds remain the oldest.

Unfortunately, all we know about *Ramapithecus* is part of the jaws and face. These, however, are of the hominid type and resemble the corresponding parts in the Dartians quite closely except that *Ramapithecus* was even smaller.

What we know of the teeth fits in very well with the

predictions based upon the Dartians. The incisors are relatively small; as a primitive characteristic, they point forwards slightly more than in the Dartians. The canines, too, are fairly small and do not jut out over the other teeth, but they are somewhat more pointed than in Dartians. The molars are comparatively large.

The tooth row is closed, without empty spaces. The cheek teeth diverge backwards as in man; the palate is strongly arched, as in man and in the Dartians.

The lower premolar is the only uncertain item. There are some jaw fragments ranging in age from early Miocene to early Pliocene, which may belong to *Ramapithecus;* in these, the anterior premolar is two-cusped, somewhat enlarged, and has an edge running obliquely forwards and outwards, and working against the upper canine.

Seen in profile, *Ramapithecus* differs from the apes of its time by its short and more vertical face with a short upper lip; all this is in decided contrast to the elongated jaws of the apes.

LATE TERTIARY APES

All the living apes are highly specialized in various ways, but as a common characteristic they have the long arms and short legs adapted to swinging hand over hand. In spite of the resemblance, the specialization arose independently in different apes. Ancestral gibbons in the later Tertiary still had short arms and long legs, and the same is true for the ancestors of the larger apes.

Could any of these late Tertiary (Miocene-Pliocene) apes have anything to do with man's ancestry? Let us make a brief survey of the main groups known.

Most of the apes in the Miocene and Pliocene, with ages ranging from about five to seven million years back to about 20 million, are now classified within the single genus *Dryopithecus*. The earliest representatives of this genus appear in Africa some 20 million years ago. At a considerably later date, perhaps about 15 million years ago, they show up in Europe and in Asia, too. We have already seen that the African *Ramapithecus* is older than the Eurasian ones, and the same seems to hold for *Dryopithecus*. This is potentially important, and I shall return to it in a later chapter.

The French savant Lartet was the first to find remains of *Dryopithecus*. His find in Saint-Gaudens more than a century ago comes from the Miocene epoch and is about 15 million years old. Various later finds in Europe, India, East Asia, and East Africa have shown that *Dryopithecus* was a wide-ranging and evidently highly successful form.

We know a great deal of its anatomy. One nearly complete skull of a small species was found by Mrs. Leakey in Kenya, and the same area has also yielded the skeleton of an entire arm. Other remains consist of jaws, teeth, skull fragments, and various other bones. Several species are known, ranging in size from small chimp-like animals to hulking creatures as large as the biggest gorillas.

Dryopithecus seems to be an ideal ancestor for the living chimps and gorillas and may also be ancestral to the orang. It is much more primitive than any of the living great apes. The arm is not particularly long; it probably was about the same length as the leg. Careful analysis of known remains indicates that *Dryopithecus* habitually moved about on four legs but was probably also able to swing hand over hand, although in a decidedly less accomplished fashion than its modern descendants. The small form, when rearing up on its hind legs, may have rested on the heel of the foot rather than on its outer side.

The hand lacks the extreme arm-swinging traits of the modern apes: the thumb is not much shortened, and the other fingers are not especially long. The hand would not have been very good at the human "precision grip" (thumb and index finger). It has been compared with the hand of the living capuchin monkeys.

The brain is relatively small—another primitive characteristic—and the face protrudes, with an almost monkey-like nasal opening rather different from that in modern apes. The teeth are quite typically ape-like with well-developed canine tusks and the telltale lower premolar. The molars have the so-called "*Dryopithecus* pattern" which recurs both in man and in the modern apes.

Man has long been considered a descendant of *Dryopithecus*. Before the true nature of *Ramapithecus* was recognized, this seemed a reasonable possibility. Now, however, we know that true hominids existed as early as 14 million years ago. What is more, they do not show any convergence towards the *Dryopithecus* type as we go back in time from the youngest (about 7 million years) to the oldest (14 million). At any rate, this would mean that the possible origin would have to be pushed back to the very earliest forms of *Dryopithecus* some 20 million years ago.

Even these early forms, however, have the specialized ape dentition. The question is whether a human dentition can be derived from it. Certainly, a reduction of the canine teeth is possible, but the peculiar premolar creates a more difficult problem. We know of no case in which a comparable specialization has been lost, once attained, by a reversal to the primitive condition. So it goes against paleontologists' grain to admit such a possibility.

Fortunately, nature itself has performed the experiment in question. We know what happens to the ape premolar when the canines are reduced and the jaws are enabled to work on each other with a rotational movement like that of man. This occurred in the extinct ape *Gigantopithecus* which evolved from *Dryopithecus* in the Pliocene. It is known from the Pliocene of India, but the main evidence comes from the early and middle Pleistocene of China.

The enormous teeth of *Gigantopithecus* were first found by G. H. R. von Koenigswald in a Chinese drugstore in the thirties. After World War II, Chinese paleontologists collected more teeth in drugstores and then finally were able to trace the giant to its lair. Fossils of *Gigantopithecus* were found in ancient caves in the province of Kwangsi in southern China. We now have a number of lower jaws of this great creature, in addition to some fifty isolated teeth. Nothing is younger than the mid-Pleistocene at the very latest—probably all of the remains of *Gigantopithecus* are in fact more than half a million years old—so that speculations about his possible relationships with the Abominable Snowman seem somewhat futile.

One striking fact about the earliest finds had been the perfectly flat wear of the cheek teeth which is typical of the human dentition. In fact, this led to the idea that *Gigantopithecus* might have been a giant man rather than an ape. But when the jaws were found, it became evident that *Gigantopithecus* was an ape after all—but quite a peculiar one.

In this ape the canine tusks had become greatly reduced in size, so that an essentially human type of chewing motion became possible. In fact, not only the canines, but also the incisors are quite small while the cheek teeth are huge. And from this it follows that the tooth rows diverge to the back. The whole arrangement does bring the Dartians to mind, especially the robust ones, although the snout was much more elongated in *Gigantopithecus.*

The peculiar traits seen in *Gigantopithecus* may well be an adaptation to a more mixed diet than that of normal apes, which are mainly vegetarian (although grain-eating habits have also been suggested). Actually, even the apes may at times hunt, kill, and eat other mammals; their main diet, however, is vegetarian. But if an ape turned to a more mixed diet, including greater amounts of meat (which it could get by hunting or by carrion-feeding), it might be useful for it to reduce the size of the interlocking canine teeth. This could be done with impunity if the ape was of the tremendous size of *Gigantopithecus;* he may have been so powerful and imposing that there was no real need for large canines in defense or attack.

Whatever the explanation, *Gigantopithecus* is an ape in which nature has performed the experiment of reducing the canine teeth. However, the premolar is still there in its original shape, utterly different from the type seen in hominids. The specialized premolar of the ape does not change back into the primitive premolar of man.

While *Dryopithecus* and *Gigantopithecus* represent the great apes of the late Tertiary, the gibbons (the genus *Pliopithecus*) were also present during this time and were found in Europe, Africa, and Asia. This form, too, is first found in Africa (together with the earliest *Dryopithecus*); it then shows up in Europe at about the same time as its greater ally.

Almost complete skeletons have been found in a fissure filling in Czechoslovakia that give us a good idea of what these ancestral gibbons looked like. They were slim, lightly built creatures somewhat smaller than the living gibbons. Like *Dryopithecus*, they had arms no longer than the legs. The enormous length of the arm in modern gibbons is quite a late development.

A curious primitive feature in *Pliopithecus* is the presence of a small tail. Modern gibbons, like other modern apes and men, are tailless.

Pliopithecus has no bearing on the origin of man, but there is a contemporary form whose bearing, if indirect, is more important and intriguing. This is the creature named *Oreopithecus*. Apart from some uncertain Russian and African finds, all of the material comes from coalfields in Italy, and dates from the very end of the Miocene—about 12–13 million years. This makes *Oreopithecus* a contemporary of *Ramapithecus*.

The earliest description dates from 1871 and was written by the French student Paul Gervais, who coined the name *Oreopithecus bambolii*—"the ape from Montebamboli." Much later, in the 1940's, the Swiss paleontologist Johannes Hürzeler resumed study of *Oreopithecus* and was soon able to pin down a number of curiously man-like traits in its teeth and jaws. Contrary to all other apes, *Oreopithecus* not only has small canine teeth but also a primitive, two-cusped lower premolar like that in man.

The lucky find of a complete skeleton in the coal mine at Baccinello in Tuscany in 1958 revealed many other remarkably man-like characteristics. The shape of the hipbone is not unlike that in man and might suggest an upright posture. The skull, too, although crushed quite flat in the skeleton that was found, shows a peculiar resemblance to some early human skulls. The face is short and vertical, without a projecting snout. The brain-case is rather rounded, and the attachment area for the nuchal musculature seems to lie very far down, almost as in the Dartians which may indicate a vertical neck on which the head was balanced. The limbs, much as in *Dryopithecus,* show a slightly longer arm, or arms and legs of equal length.

The main trouble with *Oreopithecus* is that these human traits are associated with a weird set of cheek teeth that resemble neither those of true apes nor those of men. Apes and men have the so-called "*Dryopithecus* pattern," and this is one of the most reliable indicators of a basic relationship. The cusp pattern of the *Oreopithecus* molars is so aberrant that this creature can only be distantly related to man and to the other apes. The incisors, too, have a very unusual shape.

We must conclude that the man-like characteristics of *Oreopithecus* are a result not of actual relationship, but of independent adaptation of a somewhat similar mode of life as that in ancient hominids. Such a phenomenon is called convergent evolution, and we have already seen an example of it in *Gigantopithecus,* who also evolved some man-like characteristics. Those in *Oreopithecus* seem more deep-seated, and must be due to a much longer history of hominid-like life. In fact, it probably started before the evolution of the typical ape premolar.

To us, *Oreopithecus* is peculiarly fascinating as the one kind of being whose convergence with man goes farther than that of any other primate. If *Ramapithecus* had become extinct, could *Oreopithecus* have evolved into a man? We cannot say, for extinction was the lot of *Oreopithecus*, and *Ramapithecus* got the chance.

Could *Ramapithecus* and the Dartians be regarded as convergent with hominids, instead of directly related to us? No, for in these instances all of the similar characteristics ring true. They differ from us in being more primitive and in being adapted to somewhat different modes of life, but not in the key characteristics.

Such is the record of hominid and ape evolution in the later Tertiary. But these creatures have antecedents, and we must now proceed even further back in time to take a look at them. Later we shall return to the record and see what it has to say about the life and habits of the early hominids and pseudo-hominids *(Oreopithecus)*.

THE APES
OF THE FAYUM

So far we have been discussing the later Tertiary, the Miocene, and Pliocene epochs, back to about 25 million years ago. Now we must go back to the preceding Oligocene epoch which began about 35 million years ago. There are no traces of apes or hominids in the Oligocene strata of Europe or Asia although they are rich in remains of other forms of life. We have to go to North Africa to find them, more precisely to the desert badlands at the Fayum in Egypt, about sixty miles southwest of Cairo.

The great fossil fauna of this region was known to

students nearly a century ago, but it was not until the early 20th century that the presence of primate fossils in these beds was ascertained. Since then, many expeditions have been assiduously at work here; the most recent explorations have been made by E. L. Simons.

The primate remains are found on the slopes of the Qatrani scarp, facing south and rising some 1600 feet over nearby Lake Qarun. At the base of the sequence are the remains of water-living animals—whales, sea-cows, crocodiles, and others—from the ancient Mediterranean that still flooded the area in the Eocene epoch that preceded the Oligocene. Overlying these deposits is a long series of beds with fossils of Oligocene land and freshwater animals that lived in the swampy delta of the ancient Nile of that time. The story ends with tremendous volcanic eruptions that spread a thick lava bed over the long series of deposits. The lava flow has been dated by the potassium-argon method at about 25 million years, which is the time of the Oligocene-Miocene transition.

The work of E. L. Simons, which is still in progress, has revealed an extraordinary variety of early primates in the Fayum assemblage. There are forms related to gibbons and great apes, perhaps also to the monkeys and the lemurs, and even to *Oreopithecus* and men.

One of the best-known primates from the Fayum is a rather small creature about the size of a cat. In 1911 it was given the unfortunate name *Propliopithecus* by Max Schlosser, who believed it to be a forerunner of the gibbon *Pliopithecus* discussed in the previous chapter. It is hard to understand such a notion, for there is no resemblance at all to the sabre-like canine teeth and high, sharp-edged ape premolar of the gibbons. On the contrary, the man-like traits of *Propliopithecus* are quite striking.

There are several finds of this form, including almost complete jaws and many upper teeth. The tooth row as seen in the lower jaw is closed and of an elongate parabolic shape, much like that in later hominids but gently extended from the front to the back. The incisors are small and do not jut out to the front as they do in the apes, but are more nearly vertical, as in hominids. The canine teeth are strikingly small, rather blunt and spatulate in shape. In some *Propliopitheci,* both premolars are small, rounded, rather broader than long, and two-cusped—the arrangement seen in hominids.

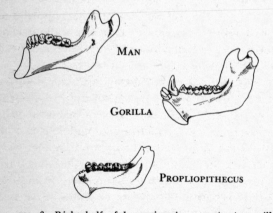

FIG. 3 Right half of lower jaw in man (top), gorilla (middle), and *Propliopithecus* (bottom), seen from the inside. The teeth, counting from the left, are two incisors, one canine tooth, two premolars, and three molars (but in the human jaw, the last molar, or wisdom tooth, has not erupted). In the fossil jaw of *Propliopithecus*, the two incisors have fallen out, so that the first preserved tooth is the small, blunt canine. Note the great canine and pointed first premolar of the gorilla.

The molar teeth are of the same type as those in men and apes, but the small size of the last of them brings to mind the hominids rather than the apes, for in the latter the hindmost molar tends to be the longest.

Seen from the side, the *Propliopithecus* jawbone is short and deep, and so contrasts with the long jaws of the apes. The chin line is more nearly vertical than the sloping ape chin, and there is a peculiar down-arching of the hind end of the jaw that may reflect the arching of the skull base so typical of hominids: it is this arching that helps to produce a vertical, rather than a protruding, face.

Is *Propliopithecus* an ancestral hominid? In its anatomy there seems to be nothing to exclude it from such a position. On the contrary, its characteristics are those we may expect in an early forebear of the Dartians and *Ramapithecus*, as soon as we shake off the compulsion to look for an ancestry among the apes.

Perhaps the only really disquieting factor at the moment is the long gap in time separating *Propliopithecus* from the earliest *Ramapithecus*. The specimens of *Propliopithecus* seem to have come from the middle part of the Fayum sequence and should be about 30 million years old while the oldest *Ramapithecus* is 14 million. It is true that reports of intervening finds of hominids from Kenya —with ages of 20 million years or more—have been made, but at present these are controversial.

The unique position of *Propliopithecus* is thrown into sharp relief by a survey of the other primates at the Fayum. Let us review them briefly on the basis of the preliminary, but well illustrated, studies available.

There are a number of primitive primates, *Parapithecus* and related forms, in which the number of premolars in each jaw half is still three, instead of two as in all other Old World monkeys, apes, and hominids. This group may be descendants of the transitional forms between the lower primates of the early Tertiary and the higher Oligocene forms. (Such lower primates, or prosimians, survive today as the lemurs, spectral tarsiers, and some other lowly forms, but the transitional group apparently became extinct before the end of the Oligocene.)

It is of interest to note that the incisors and canines of the *Parapithecus* group are quite small and the premolars, small, rounded, and two-cusped. This suggests that the smallness of the front teeth and the shape of the premolars in the hominids is quite simply a primitive characteristic that has remained unchanged ever since the higher primates came into existence.

The animals of the *Parapithecus* group are too late in time and too primitive to have any direct bearing on the origin of the hominids. One member of the group, however, has molars somewhat resembling those of *Oreopithecus,* and it has been regarded as a possible ancestor of that animal.

In the oldest strata of the Oligocene sequence at Fayum, remains of a very small, ape-like creature called *Oligopithecus* have been found. This is even smaller than *Propliopithecus,* but in sharp contrast with that form it has fang-like canine teeth and an unmistakable ape- or monkey-like premolar of the high, pointed, single-cusped type. Between the premolar and canine there is an empty space for the reception of the great upper canine.

The exact position of this animal in the history of the primates is still uncertain, but it is probably close to the ancestry of both the monkeys and the apes. Clearly, it is much too specialized to have anything to do with the ancestry of *Propliopithecus.*

In the upper Fayum, identifiable apes are already present. These include a form apparently ancestral to the gibbons, as well as the ancestral great ape *Aegyptopithecus*. Of the latter, many skeletal parts have been recovered, including an almost complete skull—the oldest ape skull known to date. Its age is about 28 million years.

Aegyptopithecus was no larger than a present-day gibbon, but it is more closely related to the great apes than to the gibbons; it is very probably the direct ancestor of *Dryopithecus*. In contrast to its descendant, it had a smaller brain and a narrow, projecting snout. The canine teeth and incisors were big, and the premolar quite ape-like. The muzzle, however, is not as broad as in the later apes, and the tooth rows diverge backwards. The molars are *Dryopithecus*-like.

The eyes already show the forward-directed position typical of monkeys, apes, and men, and the limb bones indicate adaptation to life in the trees. A tail was still present.

In many respects *Aegyptopithecus* still resembles the monkeys rather than the apes—it might even be said, as Simons notes, that we see here the teeth of an ape in the head of a monkey. Probably at this point—28 million years ago—we are still quite close to the branching point between monkeys and apes. The earliest being known without doubt to be a monkey is in fact found in the early Miocene, a few million years later.

At Fayum we must be pretty close to the origins of the stocks of apes, monkeys, and men arising from the even older and more primitive prosimians of the Eocene. Of these latter, we know numerous forms from the northern continents but so far none at all from the south —Africa or South Asia. The only trail that we can follow, at the moment, are some fragmentary specimens from the late Eocene in Burma, perhaps between 35 and 40 million years old, which appear to be transitional between prosimians and higher primates.

Even though little is known about this transition, it seems evident that the transitional forms had small canines and front teeth, and the primitive, unspecialized premolars that recur in some *Propliopithecus* and later on perhaps in *Ramapithecus*, in Dartians, and in ourselves. From forms of this primitive type the ape and monkey lines probably arose by the evolution of larger canines and the specialized premolar, all of this in adaptation to a particular type of diet. If we derive the hominids from apes, this would mean that the man-like teeth of the earliest forms would have to evolve into the ape teeth, and then evolve back into man-like teeth again. To a paleontologist familiar with evolutionary sequences, this is not a likely alternative.

Then the most logical answer suggested by the fossil evidence is this: hominids are not descended from apes, but apes may be descended from hominids.

THE FORGOTTEN EVOLUTION

Many authors have argued for a recent common origin of apes and men, and some regard men, chimpanzees, and gorillas as a very closely knit group which divided up only within the last few million years. The evidence seemed (to me, among others) quite plausible as long as we knew of no Tertiary hominids, and the Dartians were still regarded as creatures between ape and man.

One important argument comes from the comparison of blood serum proteins of different species, a method which by now has become extremely precise and subtle. Such comparisons show that the serum in man resembles that in gorillas and chimpanzees more than either resembles those of the orangutans or gibbons. This led to the conclusion that a corresponding genetic similarity exists and that it must be due to a recent common origin—a close genetic relationship.

Indeed, it has been taken to mean that the orang stock separated from the joint man-chimp-gorilla stock at a fairly early date (the gibbon stock, of course, being of even earlier extraction). Then, much later, the hominids evolved from the ancestral gorilla-chimp stock, or even, in one version, from early gorillas. (Very similar conclusions have been drawn on the basis of studies of chromosomes, their number and shapes.)

Such a claim must make us pause, for it is so utterly at variance with the fossil evidence. We have Dartians four million years back in time, showing no inclination to become more ape-like as we go back in time. Before that we have *Ramapithecus* going back to 14 million years and giving the same impression. There is something wrong here. The latest possible date of the common origin (disregarding *Propliopithecus* for the moment) is early *Dryopithecus* at about 20 million years. At that time, the migration of apes to Asia was still in the distant future, so it is almost unbelievable that the orang stock would have been differentiated; it presumably evolved in Asia, in relative isolation from the African forms. There seems to be no physical possibility that man could be more closely related to the living African apes than to the orang.

Fortunately, the dilemma can be resolved. The serological and chromosomal evidence is not really historical; it just shows present resemblance and difference but not how, when, or why they have arisen. That they reflect degrees of affinity in descent is simply a hypothesis, and it stands or falls on its compatibility with historical evidence. In the present case, the historical evidence shows that the hypothesis is untenable, and we have to look for other explanations—more about that presently.

Comparative anatomy, too, has given us a number of characteristics that seem to ally man more closely with the chimp and gorilla than with other apes. For instance, in one kind of gorilla, the mountain gorilla from the highland forests west of Lakes Tanganyika and Albert and the Virunga Mountains, the foot is rather different from the hand-like foot of other apes; it shows some resemblance to a human foot.

It is perhaps unfair to mention this characteristic at first because to regard it as evidence of a special relationship with man leads to such obviously ridiculous consequences. We would then have to regard this as the primitive type of foot in gorillas (perhaps also in chimps), dating back 20 million years in the evolutionary history of the African apes, and the hand-like foot of the chimps and the lowland gorillas as derived forms. Seeing that other apes, as well as monkeys, also have hand-like feet, such an interpretation becomes most unlikely. The foot of the mountain gorillas is obviously convergent with the human foot as an adaptation to movement on the ground rather than in the trees.

There are other and perhaps more important traits. In the gorilla, for instance, the ear looks very much like that in man, with a free lobe. The hand is broad with relatively short fingers, and thus looks more like that of man. The arms are not excessively long, the legs not very short, and young gorillas may rear up and walk about in almost human fashion, resting on the sole of the foot like man, not on its outer side. The male sexual organs, too, rather resemble those in man though their size is more modest.

These characteristics, of course, complement those other anatomical traits proving that apes in general are related to men. They show that men and apes do have a common origin. But the exact course, time, and place of evolution can never be pinned down by evidence of this kind. Only the historical evidence is capable of that, and that is the evidence given by paleontology.

It is important to stress this again because it has so often occurred that schemes of evolution based on such unhistorical (non-fossil) evidence have been proved utterly wrong. A classical example is the idea that bony fish have evolved from cartilage-skeleton fish because the relatively lowly present-day sharks and rays have a cartilaginous skeleton whereas the most advanced fish are bony. The fossil record, on the contrary, shows that the earliest known fishes were bony to a degree and that the cartilaginous skeleton of some living fish is due to a reduction of bone.

Some traits in apes and men are different. Obviously, they have been acquired after the separation of the two stocks. They are due to divergent evolution. Other characteristics are similar. We might then logically assume that they represent the inheritance from a common ancestor which must necessarily have had such characteristics.

Here the example of the mountain gorilla's foot is

sufficient to show that this is not true. It is due not to common inheritance but to convergent evolution—adaptation to a similar function.

A third possibility, equally well known and abundantly exemplified in paleontology, is parallel evolution, whereby related stocks tend to acquire similar traits independently of each other, again by adapting to similar ways of life. In parallel evolution, you might say, the resemblance tends to remain about the same although both stocks are evolving. In convergent evolution the two stocks tend to become more and more alike; in divergent evolution, less and less.

There are good examples of these modes of evolution in the history of the apes. The gibbons and large apes are divergent, but they also show parallel evolution in several respects. Both evolved various similar adaptations to swinging hand over hand—the lengthening of the arms and fingers and the reduction of the thumbs, for instance. The tail was lost in both stocks; the great apes still had a tail at the end of the Oligocene, and the gibbons had a tail in the Miocene. Also, in both groups there is a tendency to increase in size. Finally, the gorilla shows convergence with man in some respects, notably the foot of the mountain gorilla—some of the convergence is probably due to the fact that this ape is mainly ground-living.

There is nothing mysterious about parallel evolution. It can be understood readily if one considers that, given two fairly similar anatomical and genetical make-ups, the possibilities of change are rather similar for both. This should also hold for chromosomal and serological traits; the special similarities found between men and living African apes have no other possible explanation.

A trend in modern efforts to determine relationships between different organisms is to include the largest possible number of traits without giving special weight to any one of them. This is contrary to classical systematics, in which certain key characteristics were regarded as the most important. The new method has its advantages but may also create pitfalls. It operates on the assumption that all evolution is divergent, and tends to be confused whenever parallel or convergent evolution come into the picture.

For instance, if you compare men and gorillas in this way, man would appear to be slightly more closely related to the mountain gorillas than to the lowland gorillas although in fact human and gorilla ancestry separated long before the gorilla stock had split up into these two forms. I am afraid that we cannot dispense with key characteristics; fortunately, the teeth give us all the necessary ones as far as apes and men are concerned.

PART 2

THE
FUNCTION

THE
DARTIAN ARMORY

Man has been defined as the tool-making animal, and there is profound truth in this despite the discovery of tool-making and tool-usage in other creatures, particularly the apes. Tools and missiles are used on some occasions by chimps and gorillas, and there are even examples of true cultural traditions among them. In man, however, the making and use of tools and weapons has become a primary trait. Man's adaptive evolution has been dominated, for at least two million years, by his occupation with tools, and in this he is unique.

The ability to make stone tools must have been there almost two million years ago, for this is the radiometric age of the oldest tool-bearing beds in Olduvai Gorge. The earliest tools which could be said to have been fashioned deliberately consist of pebbles from which one or a few chips have been struck so that a sharp edge was formed. The actual use of this primitive chopper is not definitely known, but a reasonable guess is that it was used to help cut up dead animals for food—animals killed or found dead by the Dartians. But such an all-purpose tool would also have many other uses throughout the meal, ending with smashing the long bones to extract the marrow.

Besides the choppers, however, several other types of tools have been found in these beds. In the first place there are hammerstones and anvil stones, which were used to fashion stone tools, but there are also scrapers (with a fairly even, steeply-flaked working edge), disc-shaped tools with a cutting edge all around or nearly so, polyhedral tools with several cutting edges, and others. A number of completely unworked rocks may have been brought to the site (an ancient lake shore, now long covered by later sediments) as raw material, or for use as missiles. The remains indicate a much more varied armory than has been thought. Tools of even greater antiquity have recently been discovered in other East African deposits, but no detailed descriptions are as yet available.

The discovery of Dartian remains in association with such stone tools in Olduvai Gorge made it fully clear that these hominids were fullfledged toolmakers. As regards the South African forms, definite proof of the simultaneous existence of stone tools and Dartian remains was rather late in coming, but here, too, such association has been recorded at the Sterkfontein Type Site.

The use of other kinds of tools and weapons has, however, been documented in South Africa. As Dart pointed out long ago, baboon skulls from the Dartian caves show characteristic multilations suggesting that they were killed by a blow from a heavy bludgeon. The bludgeon could have been the lower end of the humerus (upper arm bone) of an antelope. The double ridges of the fossil antelope humeri fit the double fractures in the skulls very nicely, and the evidence has been scrutinized by experts in forensic medicine.

Evidently baboons were a favorite prey. To kill such warlike animals, the Dartians must have operated in bands. From a collection of 58 baboon skulls, Dart concluded that 47 had been killed by right-handed blows from in front, three by left-handed blows, and the remaining eight by right-handed blows from behind. The percentage of left-handedness would be about five, which is close to that in modern man. The predominance of right-handedness, then, is an ancient heritage in man.

Dart's study brings to mind a scene full of fury and movement. In a few cases the Dartians were able to club a baboon in flight, but mostly they would have been taken with their backs to the wall, fighting back in desperation.

Having concluded that the Dartians used bone tools, Professor Dart went on to suggest that the bones found in the ancient caves had all been brought in by Dartians. They would represent left-overs from their meals as well as crude weapons and tools of many kinds. The Dartians would collect the parts of their prey that had "kick" or "bite" in them. The long bones of the limbs could be used as bludgeons, or be broken up into sharp-pointed, stiletto-like weapons, or even to spoon-like scoops to help in feeding. Jaws with teeth could be used as scrapers, or if armed with pointed tusks, could be hafted to form a weapon of attack.

The idea is appealing, but there is an alternative explanation. The caves might also have been used as lairs by carnivores, and part or all of the bones may have been brought in by them. In fact, large numbers of hyena and leopard bones were found in several of the caves. Dart chose to explain this possibility away as a "hyena myth" and stated, on the basis of a number of studies of hyena dens in the vicinity of the Kruger National Park in the Transvaal, that hyenas do not collect bones in their lairs.

This contradicts the observations made in 1823 by William Buckland which made it clear that the enormous number of hyena and other bones in some European caves from the ice age resulted from the continuous use of the caves as lairs by hyenas. Not only did they leave their own bones there when the old individuals crept away to die, but also those of the helpless hyena cubs that had lost their mothers and finally masses of bones that were remains of their meals. Large numbers of the characteristic, chalky hyena feces were also found. To avoid the conclusion that the hyenas dirtied their own lairs, we may perhaps suggest that these were still in the gut of the hyena at death.

Similar observations were repeated by many later investigators. Before World War II, H. Zapfe of Vienna decided to experiment by feeding bones to hyenas and other carnivores in the Vienna Zoo, in order to find out what the bone remains would look like after the meal. As might be expected, the hyenas turned out to be experts in the art of smashing bones, vastly superior to any other carnivores. What is more, their treatment of a given bone was always characteristic. They invariably broke it up in such a manner as to expose the nutritive spongy tissue and marrow. The remains of a hyena meal turned out to be easily identifiable, and eactly the same shapes were found in the hyena caves from the ice age.

The snag is that so many of these bones are splintered in such a way that sharp edges or points are produced, and they come to look very much like something fashioned intentionally by man. Furthermore, recent studies by A. J. Sutcliffe show that wild hyenas in Uganda, Tanzania, and Kenya swallow and later regurgitate pieces of bone which may become drastically affected by the stomach juices and bowel movements.

Perfectly round holes might be bored in them, and different bones could be wedged together as if by intention.

Sutcliffe also excavated the lairs of several spotted hyenas and found that they contained large numbers of bones treated in the characteristic way. The material included numerous human bones from nearby cemeteries.

Vindication of the "bone, tooth, and horn culture" of the Dartians cannot come through denial of verified facts. The proper procedure would be to compare bones found in Dartian caves with experimentally or naturally produced bone fragments of hyenas or other carnivores. In fact, a large number of the fragments from Makapansgat were shown by E. Thenius to have the typical characters of hyena-bitten bones. But there are also observations suggesting that some of the material—the clubbed baboon skulls for example—may have a different story. As the matter now stands, it would seem that both the Dartians and the carnivores were responsible for the amassing of bones in the caves of the Transvaal.

An interesting comparison between the bone debris at Makapansgat and that of a much younger site, Kalkbank, has been made by Dart and J. W. Kitching. The Kalkbank site was inhabited by men of modern type some 15,000 years ago, who were of course greatly advanced culturally over the Dartians from Makapansgat. Yet the bone fragments were of precisely the same type at both sites. Such conservatism would seem well-nigh incredible in Kalkbank man. If, on the other hand, the bones were broken by carnivores, the similarity would be natural.

I have discussed this controversy in some detail to illustrate the difficulties we get into as soon as we try to form ideas of the interaction between early man and his environment. And yet we do know much about the Dartians in comparison with what came before them.

It has even been questioned whether the Dartians used caves as lairs at all. As K. P. Oakley has pointed out, the possession of fire would be important to cave dwellers; without fire as a protection against carnivores, a cave is as much a trap as a shelter. But there is no certain evidence that the Dartians knew how to use fire: there are no hearths, no charcoal, no charred bones. It is true that free carbon was detected in a rock sample from the Makapansgat Limeworks in 1925, and this has been cited as evidence for the use of fire by the gracile Dartians whose remains were found at the site many years later. But when new analyses were made, no carbon was found, so it does not look as if the Dartians of Makapansgat used fire after all.

Does this mean they could not have used caves as lairs? Not necessarily: a cave may still be a more easily defended place of retreat than an improvised shelter like the stone huts erected by Dartians at Olduvai. Probably the protection given by fire has also been somewhat overestimated. It has been noted, for instance by R. Perry, that tigers, leopards, and hyenas have all been known to break into camps protected by fires.

IN THE LAND
OF THE DARTIANS

Did Dartians live in open country or in the woods? We have seen that their tools and weapons are those of hunters and meat-eaters, but what kind of animals did they hunt? And were there animals, too, that hunted them? Did they live to an old age, or were they usually cut off in their prime? To answer these questions we must look to the fossil record.

Whether the Dartians lived on the steppe, in the savanna, or in the forest would depend on the rainfall. There are several ways to approach the problem.

First you can look at the animal remains preserved as fossils and decide whether these creatures were denizens of the forest or whether they preferred open country. If, for instance, you find lots of hippopotamus and crocodile bones, it seems reasonable to assume that there were large bodies of water in the vicinity and at least a belt of forest bordering the lakes or streams. A wealth of antelope and other grass-eating animals, on the other hand, suggests the open plain.

If there are plant fossils, they will be helpful. One important clue is the presence of fossil pollen grains. These can sometimes be extracted in great quantities, and they may give quite a detailed picture of the plant-cover around the site.

But you can also study the deposits themselves, for they too give clues regarding the conditions under which they were laid down. Rocks formed in dry or wet conditions, for instance, differ markedly in physical characteristics. Those produced by seasonal torrential streams in a climate with alternating dry and wet seasons may also be told apart from those that develop in a uniformly well-watered climate, and so on.

In the Transvaal caves, C.K. Brain studied the properties of the sand grains found in the deposits. Among other things, he studied their angularity. In a dry climate, soils on hillsides last a long time, and the sand grains in them break down to more or less smooth shapes. If the

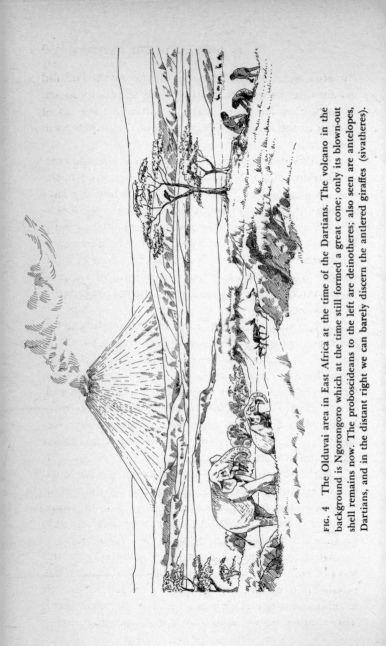

FIG. 4   The Olduvai area in East Africa at the time of the Dartians. The volcano in the background is Ngorongoro which at the time still formed a great cone; only its blown-out shell remains now. The proboscideans to the left are deinotheres; also seen are antelopes, Dartians, and in the distant right we can barely discern the antlered giraffes (sivatheres).

climate is wet, on the other hand, they are washed off and get into the caves with most of their angularity intact. Thus he was able to estimate the probable amount of rainfall in the times of the Dartians of the Sterkfontein Valley.

By this and other methods, he showed that the gracile Dartians of the Sterkfontein Type Site lived in the area in a period of rather dry climate while the later deposits of Swartkrans and Kromdraai (with robust Dartians) were formed in a time when the climate was much moister than nowadays and much of the neighborhood was probably densely wooded. So there would be a difference between the two types of Dartian: the gracile form preferred open country, while the robust Dartians were woodsmen.

If you look at the remains of fossil animals found in the cave deposit, the result is much the same. Mrs. R.F. Ewer, who studied the fossil pigs, concluded that the forms found here were rather specialized grass-feeders, so there must have been some open country around. There are no hippos, and so probably there were no large rivers. But in the wetter phase, water mongooses appear in the record, showing that there must have been a fair-sized stream in the valley.

At Makapan (with gracile Dartians), we are obviously very close to the wide open plain, for there is an enormous number of fossil antelopes of many kinds. L.H. Wells and H.B.S. Cooke say that this site probably lay in a bushy valley opening onto the nearby plains.

Does this signify a deep-seated difference in the ways of life of the two Dartian types? J.T. Robinson has pointed to a very peculiar feature in the wear of the teeth in robust Dartians which is not found in the gracile form. The enamel shows signs of slight chipping in many places. This seems to be due to sand and grit mixed in with the food. He suggests that robust Dartians were mainly vegetarians and that roots, fungi, and tubers may have been an important part of their diet.

If the robust Dartians lived mainly on vegetables, it would explain the need for great cheek teeth and enormous chewing muscles, for uncooked plant food of this kind takes a lot more chewing than red meat does. But this, of course, does not mean that they were sworn vegetarians. Probably what the robust Dartian did for a living was (to use Damon Runyon's phrase) the best he could, and this would certainly include animal food whenever available.

The sites at Olduvai Gorge, as already mentioned, represent ancient lake shores where bands of Dartians would remain for some time before moving on. Rocks might be dragged together to form rough huts or shelters that might be covered with branches. The "living floors" would be scattered with stone tools of the kind already described and with debris including waste flakes from the trimming of stone tools, remains of Dartian meals, and occasional corpses of the Dartians themselves.

Besides baboons, the Dartians hunted many other animals including tortoises, lizards, snakes, and probably various insects. Ostrich egg shells are quite common at the Olduvai sites; remains of other birds are plentiful but have not yet been studied in detail. Small mammals were taken too—hedgehogs, shrews, mice, ground squirrels, porcupines, hares, and so on (but the hares are not common, as they were probably too fleet of foot!). This sort of diet probably appealed equally to the gracile and the robust Dartians. When it came to the larger and more dangerous animals, the clumsier and less man-like robust form may have been at a disadvantage compared to his more agile cousin.

All of the evidence that has come down to us indicates that the African plains of the time sustained an incredibly rich fauna, surpassing even the East African plains of the present day in the richness and variety of their large mammals. It has been shown that grasslands of this type may carry a greater mass of living flesh per square mile than any other kind of environment, so it is hardly surprising to find the Dartians here. In fact, the distribution of stone tools from Dartian times appears to be restricted to environments of the steppe or savanna type.

There are numerous antelopes of many different kinds, among which we find ancestors of types still present as well as completely extinct forms. But there are also numerous other forms of big game.

The largest creatures here are the elephants and their relatives. Some are true elephants related to the African and Indian species of the present day; others belong to the much older mastodon group, but they were already becoming rare in Africa. These animals looked much like elephants as we know them today, but they were generally shorter-legged, and their teeth were much more primitive.

One of the most remarkable forms was the deinothere or hoe-tusker which was also elephant-like, but lacked the upper tusks typical of modern elephants. Instead, its lower jaw was armed with two great down-curving tusks looking something like an enormous hoe. The hoe-tuskers must have been impressive, yet tools found among the remains of one of these animals at Olduvai show that they were butchered by Dartians. Of course, some of the big game may originally have been killed by other carnivores and then appropriated by a band of Dartians.

The rhinos known to the Dartians are the same as the quarrelsome black and the peaceful white rhinos still present in Africa. Horses include both zebra-like forms and the now extinct, three-toed hipparions that roamed in enormous herds all over the Old World in Pliocene times. A related, but completely extinct group was the chalicotheres—great, horse-like beasts with enormous claws instead of hoofs on their feet. The heavily-armed paws probably proved fatal to more than one Dartian hunter.

Giraffes, much like those of the present day, were already present, but in addition there were enormous antlered giraffes or sivatheres (named, like many other fossil mammals from the Indian Siwaliks, for the Hindu deity Siva—the Dancer, patron of play-acting, armory, and science). In spite of their gigantic size, the sivatheres seem to have been repeatedly attacked and brought down by Dartians.

In the vicinity of lakes and rivers, hippopotami were common, while the plains, the bush, and the forest swarmed with pigs of the most varied kinds. There were pigs related to the present-day African types—the water hogs, forest hogs, and warthogs—as well as extinct types; some of these were of gigantic size and must have been dangerous adversaries.

In the midst of such an inexhaustible reservoir of large game, the Dartians themselves, no doubt, had to watch out for the large carnivores of their time. All of the most formidable enemies seem to have belonged to the cat tribe, for there is no evidence of the presence of such animals as the present-day hunting dogs which, working in large packs, can overcome most kinds of wild animals.

The great cats of Dartian times include such living forms as the leopard and lion—or at any rate animals immediately ancestral to them—as well as many extinct forms. Among these, our imagination is certainly captured by the remarkable saber-toothed cats, in which the upper canine teeth were greatly enlarged and used to stab the prey, dagger fashion. The typical form of cat in Dartian Africa is the dirk-tooth, which was the size of a leopard but much heavier. An agile Dartian might well avoid him.

More dangerous, perhaps, were the *Dinofelis* cats which were the size of a lion, but with the upper canines moderately enlarged; they were certainly more agile than the dirk-tooths and may have ambushed more than one hapless Dartian.

A large percentage of the African carnivores in Dartian times belonged to the hyena tribe, and especially to forms related to the living striped and brown hyenas. The spotted hyena, however, is also present. Perhaps the flourishing of the hyenas is, in a way, a measure of the productivity—in terms of mammalian flesh and bone—of the African savannas of those days.

A very peculiar kind of hyena present in some of the deposits would seem to be a relative of the contemporary European hunting hyena. We have more nearly complete material of the European form, and it is really a most astonishing animal. It must have been an incredibly fast runner, like the living cheetah, and probably brought its prey down by a swift dash in the same way.

The life of the Dartian, both at Olduvai and in the Transvaal, was dominated by the interaction between man and beast. Probably the foremost cause of death was an encounter with dangerous carnivores. Other hazards might be incursions of rival tribes. There is evidence suggesting that some Dartian skulls were bashed in much like those of the baboons.

At Olduvai, topped by the mighty volcano Ngorongoro, of which only the blown-out shell remains today, another danger existed. Some of the deposits from Dartian times speak eloquently of catastrophic eruptions, when the fiery cloud of superheated lava and gas that the geologist calls *nuée ardente* rushed down the slopes at incredible speed, devastating everything in its way. But the intervals between such outpourings would be too great for Dartian memories to remember from one eruption to the next.

Whatever the cause of death, Dartian lives were short. Of the gracile Dartians in the Transvaal caves, more than one third (35 per cent) died before reaching adulthood. The robust Dartians of Swartkrans and Kromdraai were even worse off: well over one half (57 per cent) of the remains are of children and youngsters.

The figures look appalling, but they are not appreciably worse than those for other early men. For instance, about 55 per cent of the Neanderthalers of Europe died before the age of twenty. So in the light of prehistoric standards in general, there is no reason to think that the Dartians did badly for themselves.

Still, the difference between the gracile and robust Dartians may be significant. The robust Dartians of Swartkrans and Kromdraai may belong to the last of their tribe, and the high mortality may indicate that the pressure of their environment was mounting. P. V. Tobias, who made these estimates of Dartian mortality, thinks that the robust Dartians of the later ages in the Sterkfontein Valley may themselves have been hunted rather than the hunters; that they were the prey of more advanced types of men. More about this possibility in a later chapter.

THE WORLD
OF RAMAPITHECUS

The beginning of the Pliocene in the Old World (12 million years ago) is marked by the immigration of a new kind of horse, the hipparion—a new offshoot from the age-long center of horse evolution, North America. Up to then Eurasia had been inhabited by small, browsing (leaf-eating) forest horses called anchitheres. They, too, originated in North America and entered the Old World in the early Miocene, about 25 million years ago. The new hipparions were somewhat larger, and their cheek teeth were well adapted for grazing, which made it possible for them to invade the great grasslands that were developing in Miocene and Pliocene times as the climate became drier. Both horse invasions probably followed the same route, a land bridge between Alaska and Siberia in the area of the present-day Bering Strait.

Although the hipparions were more advanced than the old anchitheres (and gradually crowded them out), they still retained three toes on each foot like their predecessors; modern horses, in contrast, are one-toed. Fossil remains of hipparions are very common in the Pliocene deposits of Europe, Asia, and Africa. It would seem that these small horses formed big herds rather like the zebras of the present day and mixed with great cohorts of antelopes, mastodons, and other animals on the savannas and steppes. In a way, the time of this remarkable "hipparion fauna" marks the climax of mammalian life in the Old World. The richness in species and in-

FIG. 5 A European woodland scene some ten million years ago that was possibly the environment of *Ramapithecus*. In the foreground, primitive deer (*Dicrocerus*); in the distance, okapi-like giraffids (*Samotherium*).

dividuals must have been staggering, judging from the immense amounts of fossils dating from this age unearthed in China, India, Europe, and North Africa.

But in spite of this enormous number of species and individuals, we look in vain for hominids in the savanna and steppe faunas of this age. *Ramapithecus* is systematically absent. All of the finds of this hominid are associated with the remains of a fauna of forest type.

Such is the case, for instance, in the Nagri fauna of the Siwaliks, which has recently been analyzed by I. Tattersall. Hipparions are present, but this has little bearing on the environment inasmuch as we know that these horses were equally at home in the forest and the desert. More important is the presence of typical forest-dwelling apes and other primates. There are also hoe-tuskers or deinotheres whose dentition was of the primitive browsing type, pigs and pig-like creatures called anthracotheres, woodland rhinos, and chevrotains. The Siwalik river was teeming with crocodiles. Palm trees were common. From an analysis of the sediments, similar conclusions are indicated: the area seems to have been well-forested with a few open glades, and with seasonally abundant rainfall.

The finds in southern China come from a brown coal deposit suggesting a swampy forest environment. The associated animals—unfortunately little known at the moment—include apes, pigs, and mastodons. On the other hand, *Ramapithecus* is not found in the enormously rich steppe faunas of northern China, in the provinces of Shansi, Shensi, and Kansu.

In Europe, too, *Ramapithecus* occurs in a forest environment. The fissure fillings at Melchingen carry a forest fauna of the same type as the richer assemblage of the Deinothere Sands in southern Germany. Here, the hipparions are still associated with the ancient anchitherian horses, which probably suggests that we are still very close to the beginning of the Pliocene epoch. Then there are tapirs; chalicotheres with their peculiar clawed feet; immense, browsing deinotheres; forest-dwelling pigs of many kinds and sizes; small woodland antelopes; otters, beavers and other swimming freshwater animals; and apes of the *Dryopithecus* group.

Looking finally at Fort Ternan—where the fauna still has not been studied completely—we see that all of the indications show that this is a forest assemblage too. There are both gibbons *(Pliopithecus)* and the great apes *(Dryopithecus)* and even a creature related to the swamp-dweller *Oreopithecus*. There is a woodland gazelle with low-crowned cheek teeth suggesting browsing habits, other forest antelopes, and a rhino which evidently was a browser too.

So the evidence is quite consistent. *Ramapithecus,* unlike his descendant the gracile Dartian, was a forest-dweller. What we do not know is whether *Ramapithecus* lived in the trees or on the ground although in such swampy areas as the Chinese brown coal forest, the former alternative seems more likely.

Upright, four-footed, or arm-swinging? We do not know, and we shall not know until we have more of his skeleton. It might, of course, be thought that he resembled *Oreopithecus,* who almost surely lived in the trees and who probably did a great deal of arm-swinging. But this remains to be proved.

If *Ramapithecus* lived in the trees, then the transition from a climbing life to a life on the ground took place between the stages of *Ramapithecus* and the Dartians, or in other words at some time between 12–8 and 4 million years ago. Today there is a great deal of anatomical evidence that man's ancestors were at one time arm-swinging, tree-living forms. The development of the arms, shoulders, and upper trunk so typical of men and apes is quite different from that in the monkeys, whose structure is that of a typical four-footed animal. This is almost surely due to arm-swinging adaptations in both the hominids and the apes. This is reflected not only in the skeleton but also in the muscles of the arms and trunk, as has been pointed out by S. L. Washburn. It would seem most likely that this similarity in adaptation was one of the major causes of parallelism in the evolution of hominids and apes. Granting all this, however, we cannot say as yet whether it was *Ramapithecus*, or an even earlier hominid, that practiced this sort of movement in the trees.

What about tool-using? Again, there is little in the way of direct evidence. Leakey has reported finds of antelope long bones that seem to have been smashed while fresh, as if somebody had struck them with a rock.

There is also the problem of the eoliths. These are flints that seem to have been intentionally shaped, but they are found in early Pleistocene and even Pliocene deposits and so may be much older than the earliest of the Olduvai choppers. Around the turn of the century, for example, large numbers of such eoliths were collected in the early Pliocene deposits at Aurillac in France. But later on it was shown that natural agencies might produce chipping of stones. In exceptional cases, then, random chipping might result in shapes resembling real tools, so the eoliths were discredited.

In the late fifties, however, A. Rust attempted to rehabilitate the eoliths. Studying the Aurillac specimens, he decided that about 10 per cent of the 7,000 flints collected by Westlake more than half a century earlier could only be explained as products of intentional stone-work. He found interesting similarities to eoliths of a much younger date from the so-called Heidelberg Culture (associated with the renowned fossil man of Heidelberg), dating back about half a million years. The characteristic implement here has an edge with a nose-like projection and is termed a nose-scraper. According to Rust, the same type is found at Aurillac. Moreover, Rust states, this Pliocene industry is so advanced that it must have a long history: its roots would go deep down into the Miocene. If all this is correct, *Ramapithecus* would evidently be the tool-maker in question.

But although these possibilities cannot be summarily rejected, it must be admitted that most authors are reluctant to follow Rust's lead. To them, both the Heidelberg Culture and the Aurillac eoliths are nothing but randomly chipped stones formed by natural agents. And it seems reasonable to say that, before we can accept the eoliths as hominid-made objects, we must at least have evidence of their presence in the same strata with bona fide hominid fossils.

BEGINNINGS
IN AFRICA

If the environment of *Ramapithecus* is fairly well known, surprisingly little can at the moment be said about that of the possible Oligocene hominid, *Propliopi-*

thecus. To be sure, there is a rich fauna in the Fayum. But most of the fossil remains come from two well-defined beds containing large numbers of petrified tree trunks that were once stranded here when the proto-Nile was laying down the deposits that now form the Jebel Qatrani. There is a very thick lower fossil wood zone containing many remains of animals from the early Oligocene—including tiny *Oligopithecus,* the earliest ape-monkey or monkey-ape. Then there is a thinner upper fossil wood zone which has yielded the bones of the ancestral great ape, *Aegyptopithecus.*

From the preliminary results, however, it would seem that at least the majority of the *Propliopithecus* remains were found in neither of the fossil wood zones, but in the intervening, mid-Oligocene levels. So it is just possible that this small early hominid, if such it is, inhabited a kind of environment different from that of the Fayum apes. The latter were almost surely forest-dwellers. What, then, of *Propliopithecus*? There is a good chance that we shall know one day.

The presence of all of these higher primates in Africa at a time when nothing of the kind is known from other continents assumes a special significance when combined with what we know of the geography of the past. To understand it we have to go into some concepts of modern geology. Again, we are dealing with results and theories of the last few years.

It is now thought that the earth's crust consists of a number of rigid plates of varying sizes and thicknesses, moving slowly in relation to each other. The force generating the movement is thought to be brought about by convection currents in the layer underlying the crust, called the mantle. These slow currents transport heat from the molten core of the earth towards the surface.

There is no shortage of heat, for the store is constantly being replenished by the radioactivity of some of the elements that make up the earth.

The plates move quite slowly, at the rate of one or a few centimeters a year. Yet in geological time the effects are great. Whenever two plates move away from each other, a rift is opened between them, and this is filled by material rising upwards from the earth's mantle. If this happens at the bottom of the sea, we get a structure like the mid-Atlantic ridge with its internal rift valley. On land, the East African and Jordan system of rift valleys is an example. Later, the rift opens up a seaway, as exemplified by the Red Sea.

When the plates move towards each other, there are two possibilities. Either one of the plates is thrust beneath the other and sinks down to become part of the mantle again, thus forming a deep-sea trench in the area, or else the shock of compaction is taken by the crustal plates themselves, and they are folded into a mountain chain.

There is also the possibility that the two plates will slip past each other in opposite directions, as they do, for instance, along the famous San Andreas fault line in California. Here, the movement is jerky: a stress is built up and then relieved by an earthquake in which the two plates slide against each other.

At some point in time more than a hundred million years ago, the Atlantic started as another rift. It has now been clearly demonstrated that in the geological past, North America was joined to Eurasia while the southern continents—South America, Africa, the Indian peninsula, Australia, and Antarctica—formed another supercontinent. The two supercontinents were separated by an ocean, the Tethys Sea, which opened up to the east but narrowed to the west where land bridges may have joined the northern and southern supercontinents.

Such was the situation as the great reptilian fauna of the Jurassic and Cretaceous periods first evolved, and as there were no difficult migration barriers, the fauna was much the same all over the world. We find roughly the same kinds of dinosaurs, for instance, in Africa and North America.

At the beginning of the Age of Mammals, when the mammals embarked upon their great evolution and diversification, the situation was extremely different. At that time, rifting of the old supercontinents had proceeded far enough to produce a number of isolated or almost isolated land masses, each of which became the scene of an evolutionary drama of its own. One such chip of the ancient southern supercontinent was Africa which was nearly isolated from the other continents in the early Tertiary. In Oligocene times, it was still dominated largely by forms of local origin, even though a few immigrants from other parts of the world had entered.

Now we can see the meaning of the rich fauna of higher primates in the Oligocene of North Africa, while such creatures are unknown in other parts of the world. Clearly, Africa was the original center of evolution for the Old World monkeys, apes, and early hominids; just as South America was that of the sloths and anteaters, North America that of the horses, Australia that of the kangeroos, and so on.

There are several other local products in the Oligocene fauna of Africa. Besides the primates, the most important group is that of the proboscideans—the order that now survives as the two living species of elephant. At Fayum we find the earliest known mastodons, still very primitive and of moderate size—with a shoulder

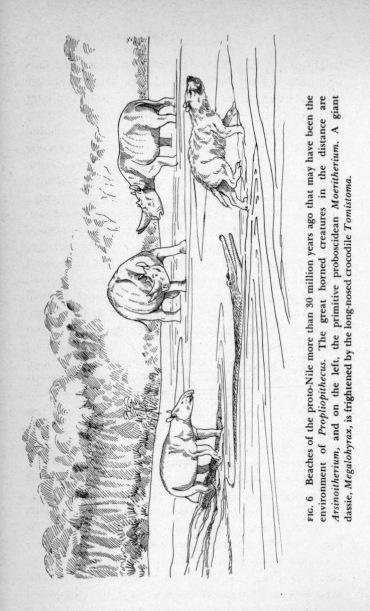

FIG. 6 Beaches of the proto-Nile more than 30 million years ago that may have been the environment of *Propliopithecus*. The great horned creatures in the distance are *Arsinoitherium*, and on the left, the primitive proboscidean *Moeritherium*. A giant dassie, *Megalohyrax*, is frightened by the long-nosed crocodile *Tomistoma*.

height of 4–5 feet. They have peculiar, elongated jaws with tusks in both the upper and lower jaws. There are also ancestors of the deinotherian hoe-tuskers of later days, as well as a short-lived, amphibious branch of early proboscideans, whose mode of life may have resembled that of the hippopotamus of the present day. Outside Africa, there is no trace whatever of proboscideans from the Oligocene, so we must conclude that these mammals originated in Africa.

The largest mammals of the Fayum are not the proboscideans but the extinct arsinoitheres, gigantic horned ungulates with a distant resemblance to the rhinoceroses. They died out soon afterwards, but the related conies or dassies survive in our time. These are small, hoofed animals; some species live in colonies in small caves and burrows while others inhabit hollow trees. Some of the Oligocene conies were much larger and reached almost the size of a tapir. Like the proboscideans, arsinoitheres and conies were of African origin, as were some of the large, wingless, ground-living birds whose remains are found in the same layers.

There are also some immigrants from Eurasia: a few anthracotheres and other pig-like forms, some rodents, and a number of primitive carnivores. But the great majority of the typical Eurasian land animals of the Oligocene are missing—all of the odd-toed and most of the even-toed ungulates, the hares, the higher carnivores, and most of the rodents and insectivores. This shows that Africa must have been almost isolated. At most there was a tenuous contact across a narrow land bridge, or a string of islands, or possibly across a desert belt. In any case, there was little intermigration of land mammals between Eurasia and Africa in the Oligocene.

In the Miocene and Pliocene, on the other hand, a

massive exchange of land animals took place. This is the time when proboscideans, apes, and finally monkeys and hominids spread out of Africa to conquer new areas in Europe and Asia. At the same time, Eurasian animals of many kinds pressed into Africa, causing much of its original population (such as the arsinoitheres and giant conies) to become extinct.

Thus, Africa is the land in which the Hominidae originated. Their rise, evolution, and migration in space and time are part of the grand process that is gradually changing the geography of the world, that breaks old contacts and forms new ones, that makes continents into drifting Noah's arks and sends them towards new landfalls in an ever-changing environment.

THE
SHADOW LAND

So much has been written about the transition from a life in the trees to an upright, two-legged existence on the ground that we might be tempted to visualize it as a sort of shadow land traversed in a long, mad rush by the prospective hominids who finally emerged in a garden of green grass and brightly shining sun. It is seen as a phase of evolution when they were "particularly vulnerable to predators and other hazards of the environment." There could be several reasons for holding this view. One would be that erect posture (necessary for running and walking) was not yet as efficient as that of the modern hominids. Another would be that the canines had shrunk and were not adequate for attack or defense. The hands, too, were still primitive, and only a few tools, of the crudest kind, were made. The brain was still comparatively underdeveloped.

Although theoretically possible, this hypothetical situation must be regarded as extremely unlikely. We ought rather to postulate that unless really compelling evidence to the contrary can be brought forth, the ancestral hominids were well adapted to their mode of life and led a highly successful existence. This was also probably true for the transition from a life in the trees to a life on the ground, and from a four-legged to a two-legged posture. The logic behind the assumption is obvious. A population that is poorly adapted to its current way of life gets very little chance to evolve into something else (for example, into man)—it is much more likely to become extinct.

Of course, it might happen that the number of individuals in the ancestral form shrank until the random processes of genetic drift supplanted the orderly process of selection that governs every population of adequate size. Such a random change would then lead to a situation that was, somehow, pre-adaptive for a new mode of life—that of early man. This is a possibility, but it is so full of remarkable coincidences that it reads a little like an Ellery Queen plot.

Appearances may be deceptive, and the "incompleteness" of Dartian adaptation is a case in point. Compared with ourselves, the Dartians don't seem to have adapted very well. But the shortcomings of the Dartians seem few indeed when compared to those of one of the great sauropod dinosaurs of the Mesozoic era, such as the well-known *Brontosaurus*. That 30-ton body resting on pillar-like legs, the little head on a long neck with only a few feeble teeth in the mouth and a ridiculously small brain, that long tail, apparently the sole defensive weapon of the silly creature—do they not look utterly inadaptive, doomed to speedy extinction? And yet they were not, for dinosaurs of this general type persisted and were present all over the world for more than 120 million years. No matter how weird and unlikely they seem to us, they must have been superbly adapted to their own environment.

I believe we have to assume that not only the Dartians, but also their predecessors, including *Ramapithecus* and *Propliopithecus,* were efficient, well-integrated organisms, doing very well for themselves. While the range of *Propliopithecus* was restricted by geographic barriers, *Ramapithecus* ranged from Europe in the west, to China in the east, so that it was obviously a highly successful creature. These early hominids must have had a distinctive mode of life diverging notably from that of any present-day apes or monkeys, and also, though in a different way, from that of later hominids.

Transitions from one major life zone to another involve continuously well-adapted populations in normal cases. For instance, the transition from fish to amphibian has been discussed by A. S. Romer. He suggests that fish evolved legs, paradoxically, to be able to stay in the water. They used them in moving overland in quest of new water basins when those that they inhabited had been reduced by drought. The ancestral fishes had strong, flexible fins that were probably excellent as a support when moving on land. Gradually, they evolved into true limbs while the fish remained on land more and more of the time. Actually, there are several kinds of fish today that move about on land and even climb trees, and if there were no other land animals present to compete with them, they might very well give rise to a new stock of land-living vertebrates.

The converse transition—from life on land to life in the water—is beautifully exemplified by many living mammals in which we can study every stage, so to speak, of the process. An early stage is represented by the polar bear, a powerful swimmer which trails its prey (seals) in the water but hunts only on land or on the ice. This existence seems to be a no man's land between two major life zones, and the polar bear thrives in it. This does not exclude the possibility that in the distant future a polar bear population might take on the habit of hunting and eating in the water; this would presumably initiate a far-reaching remodelling of the limbs. At present, the limbs of the polar bear have changed very little, as they have to be used for hunting (on the ground) as well as for swimming.

Successively more advanced swimming adaptations are seen in the otters (which take their prey in the water), the sea lions, the seals, and the whales.

In an analogous way, the development of flying structures may begin with climbing and jumping forms such as the ordinary squirrel. The next step would then be the development of a skin fold, thus producing the gliding forms as seen in the flying squirrels, colugos, and certain lizards. Perhaps the most interesting examples are some small jumping frogs whose webbed feet form "wings." Here, an increase in the length of the toes and the size of the skin fold would lead to the evolution of a flying wing much like that of the bats.

Final stages in this evolution are completed with the wings of true flyers like the bats, birds, and certain extinct flying reptiles called pterosaurs. Once they reached this stage, they tended to persist unchanged. That bat wing has been the same for 50 million years. The bird wing dates back more than 100 million years, and the pterosaurian wing, similarly, was in existence for more than 100 million years.

The point is, however, that not only the "finished" types, but also various intermediate stages may persist, showing that they, too, are advantageous as such. There are, for instance, colugos in the earliest Tertiary period, nearly 60 million years ago.

The transition from life in the trees to life on the ground probably took place in a similar manner. The habits of the early hominids probably changed gradually, so that the time spent on the ground tended to increase slowly. But for a long time the branches and foliage meant home, security, and resting place.

This similiarity to the transition from water to land may be quite close. W. W. Howells, W. F. LeGros Clark, and others suggest that early hominids descended to the ground merely to be able to move to other trees that were too distant to be reached by jumping. In this way they could make a living in the more open gallery woods, and finally on the savanna where trees stand in isolated spinneys. Later on, caves and man-made shelters assume the role of the ancient tree house.

Exactly when did this occur? We cannot say for certain, but at this point I should come out squarely with my opinion.

It seems probable that *Ramapithecus* was mainly a tree-living form: otherwise its presence in swampy forest surroundings like that of Keiyuan would be difficult to understand. Perhaps, then, the change from the trees to the ground may have occurred within or just after the *Ramapithecus* stage. If this is correct, the time involved would be the earlier part of the Pliocene epoch—approximately 12 to 7 million years ago.

This is simply my personal reading of the evidence now in hand, and I do not want to be dogmatic about it. The change may have occurred earlier. It is not likely to have occurred later.

THE ENEMY

Carnivorous mammals were a major hazard to the early hominids who came down to the ground. Perhaps, as G. Kurth has suggested, one reason for aggressiveness in man may be sought in the early confrontations between attacking carnivores and suprised hominids. The band that arose in immediate and fierce defense might survive where the more timid groups would be decimated. This would be natural selection, favoring such genetic factors that make for a tough, ready-to-fight disposition. The idea is interesting, but its worth depends almost as much upon the type of carnivore involved as on the kind of defense put up. Wolves and hunting dogs, for instance, do not seem to be intimidated by any display of fierceness. Hunting dogs attack and finally overpower tigers even though the dogs themselves may be killed in the fight; wolves will bring down a moose even if it pounds one of them to death with its hooves. Other carnivores scare more easily.

The important point is that neither wolves nor hunting dogs were present in the environment of *Ramapi-thecus*. When discussing the dangers that awaited early man as he descended from the trees, we seem to take it for granted that the carnivores were more or less the same as the ones present today in Africa and Eurasia. We tend to think of the lions, leopards, and hunting dogs of Africa, and of the tigers, leopards, wolves, and dholes of Asia. But all of these are comparatively late arrivals. Both the African hunting dogs and the dholes of Eurasia (this red dog was present in Europe during the ice age) evolved from wolf-like ancestors within the last two or three million years. The wolf and dog genus *Canis* itself evolved in North America and did not enter Eurasia until the middle Pliocene—it became really common only in Pleistocene times. The great cats of the modern type—tigers, lions, and leopards—are also of quite late date. In the Pliocene their ancestors were still of a size little larger than that of a kitten.

Let us take a look at the carnivores of the early Pliocene. The Eurasian ones are the best known, but discoveries are also being made in Africa.

There are a few creodonts, primitive carnivores of a type that flourished in the early Tertiary, which survived to Pliocene times in India. But the great majority belong to the modern families of carnivores. Still, although we can identify them as members of the cat family, the dog family, the bear family, and so on, it must be noted that all of the more powerful forms that may really have been dangerous to early hominids belong to genera that are now extinct. This means we have no detailed knowledge about their appearance and habits.

There were of course many kinds of very small car-nivores that would present no real danger to the homi-nids, but which might become quite a nuisance, for example by stealing a kill the moment our ancestor's back was turned. It is in this group that we place members of the mustelid family—weasels, otters, martens, badgers, and their extinct relatives. A few forms—for example, a primitive wolverine and a member of the honey badger group—attained the rather respectable size of their present-day descendants, but even these large mustelids were probably almost harmless to the hominids. The same is true of the civets and the small, civet-like hyaenid *Ictitherium,* which is quite common in the fossil record.

Then, on the other hand, there are the very large and probably heavy carnivores which, even if invincible as such, could be spotted at a distance and avoided. Among these animals are the gigantic bear-like dogs, some primitive but equally big true bears, and the great saber-toothed cats of the *Machairodus* group. Some of these may well have been regarded as benefactors rather than enemies. The big *Machairodus* probably preyed on large game, which it killed with its powerful paws and by a stabbing blow with its long, sharp, scimitar-like canine teeth. Its teeth, however, are such that it could live only on soft red meat and entrails. So there would be a lot of food left, and the early hominids were certainly smart enough to notice this.

When we come to the range of medium-sized, fleet-footed carnivores, we are probably dealing with some of the arch-enemies of the early ground-living hominids. Looking first at the dogs, we note that even though true *Canis* is absent or rare, there were some highly predaceous members of this family in the early Pliocene. This is shown by scanty remains of short-faced dogs with a peculiarly cat-like dentition. It seems clear that these "felinoid" dogs were almost exclusively carnivorous in habit, in contrast with most present-day dogs which are almost as omnivorous as man. Unfortunately, very little is known about these creatures. The meagerness of the record suggests that these animals were not particularly common, and so far they have been found only in Europe and North Africa.

A second group is that of the medium-sized cats, which could probably pursue their prey in the trees as well as on the ground. They belong to an extinct genus, *Pseudaelurus*, but were probably very similar to modern cats. They varied from the size of a bobcat to that of a leopard, and at least the larger ones may well have been a real menace. The same may be true of a relatively small and agile type of saber-toothed cat whose remains, like those of *Pseudaelurus*, have been found in both Europe and Asia.

The role played today by animals that hunt in packs —wolves and hunting dogs—was then played by a quite different group of carnivores. These were certain lightly built hyenas, of which are known several species belonging to the genus *Palhyaena*. Fossil remains of these animals are very common and almost outnumber those of all other carnivores. At some sites it actually seems as if a pack of palhyaenas had been caught in a catastrophe such as a flood. There is every reason to think that they were the most common carnivores on the savannas and steppes of the early Pliocene. They range from the size of a terrier to that of a wolf, and their teeth would indicate that their habits were highly carnivorous and predatory. Encounters between early hominids and palhyaena packs may have been common, and such encounters may have done much to promote intragroup cooperation and loyalty, and intergroup aggressiveness, which are such pervading characteristics in man.

A WAY OF
LIFE IN THE TREES

It has been said that the great naturalist Cuvier was able to reconstruct a whole animal from a single bone or tooth. In reality, no scientist can perform such a miracle. What a competent osteologist can do, though, is to identify a bone belonging to an animal already known and then say what the animal looked like. But he cannot reconstruct a hitherto unknown animal on the basis of a single tooth or even the whole tooth row. He can determine only whether the unknown creature was related to some other animal already known to science and how close the relationship was. This is the situation with *Ramapithecus*. We know that it was a hominid back in the later Tertiary, but we do not know what it looked like or how it lived.

And yet it is legitimate to speculate about the ways in which our ancestors in the Miocene and Pliocene lived and about their appearance. So far we have kept very close to the actual record: we will now speculate. There are good precedents—much thought and speculation have gone into attempts to visualize a coherent story of man's Tertiary antecedents. Some of the most important contributions have been made in recent articles and books by C. F. Hockett, R. Ascher, D. Morris and R. Ardrey; these and other discussants come from many different fields of study, and the synthesis of their approaches has been particularly fruitful and stimulating.

It is my intention here to paraphrase some lines of thought and to give a personal outline of what I think may have happened.

Our tree-living ancestors perhaps 15 million years ago were creatures about the size of a five-year-old child. They were furred (a naked skin bruises too easily in the trees); we would probably have called them apes, yet they lacked the projecting snouts and big eye-teeth seen in many apes. Their faces were probably more expressive than those of apes and may have had an "ape-baby" quality. The eyes looked straight forwards, with stereoscopic vision; the ears were like those in man. The lips were thin and the mouth large: the nostrils, over a short upper lip, faced somewhat downwards rather than forwards.

The arms were relatively long, and they moved easily in the trees, on all fours, balancing upright on branches or swinging by the arms. At times they would descend to the ground where they could run rather quickly in an upright position, but they were not very good at walking slowly on two legs. They would be in motion only in the daytime; they would sleep in the trees and also take to the trees for shelter.

They lived mainly on insects, grubs, eggs, young birds, and other small animals, as well as fruits, nuts, berries, and other vegetables. But they would also be on the alert for carcasses lying on the ground, perhaps killed by carnivores, and they would try to snatch away meaty bones and other morsels if they got the chance.

They lived in small bands of perhaps ten to fifty individuals led by one or a few males. Such a band would have a more or less clearly delimited home territory from which other bands were fended off. Threatening display against intruders would consist of waving and perhaps

throwing missiles like sticks or stones; they were probably in the habit of carrying such things about—in their
mouths or in their hands or feet, depending on how they
were moving. There was probably little baring and
gnashing of teeth, for their canines were not very impressive. Instead, screaming and chattering would be more
important.

The females would have a monthly sexual cycle.
Mating may still have been polygamous at this stage. The
position in copulation may have varied, as in present-day
apes, although mounting from behind may have been
prevalent. Face-to-face matings may also have been common. In general, one young was born at a time. Twins
would be at a disadvantage, being cumbersome to the
mother. The young were helpless and needed a long
period of care. They were suckled at breasts which were
not nearly so well developed as those of a modern
woman. Children reached puberty at about eight years of
age and stopped growing at ten. With luck, a few individuals lived to thirty or more, but most died much
earlier from diseases, accidents, or other causes.

They communicated by a system of calls, gestures,
and facial expressions, the understanding of which was
innate. Originally, such calls and gestures represented
simple messages like "food," "danger," "keep off me,"
etc. In time, perhaps, some of them were blended or
combined to express more complex situations. As long
as the calls were understood innately, rather than by
learning, the system remained rigid and lacked some
characteristics of true language.

One of the characteristics of language is that new
things may be said and understood—linguists term this
"openness" and contrast it with the "closed" nature of
a call system in which innovation takes place only by the

slow process of biological evolution. It will take many thousand years to add a new call to the vocabulary. Still, evolution of that kind does take place; we know, for instance, that the call system of gibbons has differentiated into local "dialects."

Another characteristic of true language is that we can speak of things that are out of sight or in the past or the future, while a call system is always concerned with the here and now. Also, a language consists of elements that have no meaning in themselves but are combined to form words and sentences. In the call system, on the other hand, the sounds in the call colored the utterance entirely, just as if we were to interpret any word containing an "s" as an angry hiss.

Finally, a language has to be learned. We are born with a capacity to learn languages—any language—but not with an innate understanding of it. This capacity has been related by some to our ability to correlate different modalities: for instance, in the dark we can identify, by feeling, cut-out shapes like squares and triangles and, without effort, correlate them with the corresponding visible shapes, but this is a stumbling block to even the most intelligent animals.

No one knows when the call system evolved into a true language, but there is some reason to suspect that it happened after the hominids had descended to the ground—more about this in the next chapter.

In any case, it seems probable that *Ramapithecus* had a varied vocal repertory and this has played an important part in the life of early hominids. The arching of the palate indicates a well-developed tongue rather than the flat tongue of the apes, and this is necessary for articulated speech.

The use of standard calls gradually became coupled with the existence of simple thought processes. The origin of thought has been discussed recently by ethologists and has been correlated with behavior of the type leading to some of the typical instinctive actions of the species in question—for instance, nest building, mating, eating, and so on. Such preliminary behavior, also called appetitive behavior, is very common in our own world; in fact, most of the things we do really come under this heading—working for money, preparing food for dinner, dressing up for the pleasures of a party, or stripping for the pleasures of the bed. With the consummation of the appropriate action, however, the chain ends, and some time will lapse before a new series of actions towards a similar end is initiated.

It is at this time, while the level of excitation is still quite low, that idle attempts at appetitive behavior may occur and may perhaps take the form of an "inner rehearsal" for the coming movements and actions. That something like this may actually be taking place has been shown by measurement of the electric potentials in the central nervous system. This would then be a kind of primordial thought process, and, as P. Palmgren has pointed out, it might assume special importance in cases where different alternative actions present themselves with about equal attraction. The jackass which must choose between two equally attractive wisps of hay would illustrate the situation.

In this way, then, we might perhaps try to visualize the early, tree-living hominids of the late Miocene. As may be noted, we can speak only of generalities. Nothing can be said at present about special, specific adaptations and habits, although we may feel quite sure that we would be struck by many such traits if we were privileged to go back in time and see our ancestors in the flesh.

I have even left open the question of whether they still had a tail. They probably didn't, but we cannot be quite sure. The gibbons of that time did, and yet they are now as tailless as we ourselves are.

A WAY OF
LIFE ON THE GROUND

According to the classical theory, the hominids had to shift from the trees to the ground because forests retreated and grasslands spread with the increasing dryness of the late Miocene and Pliocene. There are many permutations of the theory, but the basic idea is that the thinning-out of the forest forced the hominids down to the ground for shorter or longer spells; thus, they had no choice but to adapt to this new way of life or become extinct.

I believe that this basic idea is wrong. It would imply a non-adaptive situation. Had this been the case, the tree-living hominids would simply have become extinct where the forest vanished, and would have survived where the forest did.

The true explanation must be that the hominds came down from the trees because they had reached a stage when a lucrative new mode of life was available on the ground, one which was at least as good as the old life in the trees. With the spread of grasslands, great areas suitable for this way of life were opened up. With the usual opportunism of evolution, the hominid stock seized upon this chance. The ancestral baboons did likewise at about the same time, but there was no competition between the two groups. The baboons became ground-living, four-legged herbivores; the hominids became ground-living, two-legged carnivores. They were no dismal exiles from a vanishing Garden of Eden; on the contrary, they were pioneers conquering a new zone of life.

It is here suggested that this process—invading the open glades and, later, the savannas and steppes—began in the late Miocene or early Pliocene and that it was in full swing about the middle of the latter epoch some five or six million years ago.

From their arm-swinging habits the hominids were already accustomed to an upright posture and at an early stage they were good at running, partly because they were so small and lightweight. Now they became more proficient in walking as well. Carrying things was still much in fashion, and the hands were now free to take over this role entirely except when climbing trees. At the same time the feet were losing their grasping function and could adapt to walking and running.

The hominids still sought shelter at night in the trees, or perhaps in cliff areas among the rocks, but most of the waking time was spent on the ground. Much more game is available here than in the trees, and as the animals are bigger, a kill is more rewarding.

Let us visualize a band of hominids trekking in search of new hunting grounds. Walking upright, they are tall enough to see over the savanna grass and they keep a close watch in all directions for game and enemies. If there is danger and a tree is within reach, the band will run there; the slowest and most awkward runner will lag behind and may become a victim. If there are no trees, safety may lie in hiding or in fierce collective defense.

The band may be a tribe or a family—the tribe tends to be bigger, the family smaller, than the tree-living band. As the tribe or family moves along, they carry with them babies and food as well as weapons. A hunting party will be all-male or include at most a few childless females, and only weapons will be carried.

A rabbit-like creature is sighted, and suddenly there is a swift rush. Panting, the victorious hunter returns with his prey. After that explosion of movement, sweat drops glisten in his thin fur. The thinner the fur, the less the danger of suffering heatstroke from inability to dissipate the heat accumulated during the dash. So there is selection against a thick fleece and for an increase in the number and size of sweat glands. This is a significant ground-living innovation. Tree-dwellers do not have efficient sweat glands. But there will soon be additional selection for a naked skin.

The pseudo-rabbit is soon slit open with the teeth, fingers, and sharp pieces of rock. When it is eaten, the inside-out skin may be lugged along. You could chew on it later when you felt hungry again. There is a glimpse of forethought here, occasionally, and again it is grist to the mill of natural selection. The chewing makes the skin more pliable, and somebody may discover that it can be used as a bag, or to keep warm in the cold night of the savanna.

But such treks are hard on the small children and on women carrying babies in their arms, and so the band makes itself a semi-permanent home in a suitable area, perhaps preferably wooded. Here the children and the parents tending them will remain while the hunting parties go out to forage. Thus the family bond becomes most important. With children needing such a long period of care, parents often have to divide up the day's work between them. It is now necessary that the foraging parent faithfully perform his duty of bringing back food to the stay-at-homes. While in the past only the mother-child bond had been significant, that between father and child and, most important, the sexual bond between the parents, evolve. Selection, as always, takes a hand, to produce the man and woman a million generations of the opposite sex will dream about.

Sexuality gradually pervades the life of the hominid, in great contrast to its incidental role among other primates. The two-parent family becomes the nucleus of the population. The division of labor is usually envisaged as complete, with hunting males and baby-sitting females. Such a division, however, tends to produce great sexual dimorphism in size, the males becoming very much larger than the females. There is actually no great size difference, and so the division of activities probably was not that complete.

Face-to-face mating becomes increasingly common and contributes to the development of affectionate feelings of a kind that earlier existed only between mothers and their babies. At the same time, various bodily characteristics already brought about in adaptation are seized upon as sex signals. The already thin fur becomes even thinner, for the intimacy of the naked skin is greater than that of the furred. Lips become fuller and female breasts enlarged—in the opinion of D. Morris, as a mock-up of the visual signal effected by the labia and buttocks in the ancient breech presentation. If this is

correct, it may suggest that face-to-face copulation became really common only well after the hominids had
become two-legged, ground-living creatures, for it is
only in this phase that the gluteal musculature develops
to form the typical human buttock. The near universality
of the skyfather-earthmother myth testifies to the importance, in most human populations, of this coital position.

While hair was lost over most of the body, head hair
probably grew thicker for protection against the sun-
shine. Genital and armpit hair, as well as the beard in
males, was retained as a sexual signal and possibly as a
scent trap. The beard also became a signal of a dominant
position in the male hierarchy. The genitalia themselves
are strongly affected. The female clitoris develops to a
size and sensitivity that gives her the capacity for orgasm,
unique among female primates, and there is a correlative
increase in size of the penis, increasing the massaging
effect in coitus.

At the same time territorial behavior becomes increasingly important and complex. The tribe forms a
territorial unit guarding its own hunting grounds and
homes. Relationships to neighboring tribes range from
peaceful cooperation to life and death struggle. A high
selective premium is placed on faithful comradeship and
unswerving loyalty in the face of external enemies or
hazards.

The family forms a territorial sub-unit kept together
by the family bond. Here is the seed of possible conflict
between family loyalty and tribe loyalty. To avoid disrup-
tion, rituals and taboos develop. Cultural evolution
is on its way, interdigitating with biological evolution.
The hominids are gradually becoming men.

There is also a tendency towards increasing body size. With larger size, the individual becomes stronger and is able to overpower larger game and to defend himself and the tribe against larger carnivores. But at the same time the size increase makes greater demands on his limbs. The earliest ground-living hominids were so small and light-footed that their "unfinished" two-legged adaptation did not really matter; the mechanical stresses remained moderate. The adaptation evolved as they grew in size and weight. Legs became longer, feet less flexible, and the heel and big toe combined to form the arch of the foot. Similarly, the adaptations of the hip girdle and the muscles were perfected.

The hands, now free to manipulate weapons and other objects and satisfy the ever-growing curiosity of the brain, became more and more skilled at doing so. And with the skill of the hands, that of the brain is augmented. Weapons and tools are fashioned; the idea of stonework flares up and spreads like wildfire. Man the tool-maker is coming into his own.

Again, the ancient call system is being superseded by a true language, though at first with a very limited vocabulary. (It is often said that so-called primitive languages of our days have an especially rich vocabulary—but they are not truly primitive in the sense that they did not recently arise from a call system.) We do not know if language evolved *from* a call system or *beside* it. The human ear is most sensitive to sounds with a pitch corresponding to the frequency of about 3000 cycles per second, but the ordinary speaking voice of men, women, and children is at least two octaves lower in the scale. The call system is likely to have had the higher pitch. Was this then a real change, by which the high-frequency channel was left for emergency screams? Or did the ordinary speech arise from low-pitched chatter at the speaking-voice level rather than evolve out of the call system? In the latter case, the two methods of communication might have remained in operation side by side for some time until the spoken language was sufficiently rich to take over entirely. Use of the calls would gradually die away, and the understanding of them would slowly become garbled in our genetic make-up, due to random changes, so that we now get little more out of an inarticulate scream than a sense of desperate urgency.

Our hominids are now big and accomplished enough to become less inclined to climb trees for the night. They turn to caves or rock shelters, or fashion their own huts. They are now approximately at the stage of the Olduvai Dartians, and so we have come full circle in our narrative of early hominid evolution.

And at this point new evolutionary vistas are opened. The invention of language and the large-scale manufacture of durable stone tools are the starting point for a new kind of evolution in which the cultural evolution becomes inextricably combined with the genetic. In this situation, brainpower becomes more important than ever before. The new modes of life opening up for hominids could be exploited fully only by an increase in intelligence, and in fact it is from the Dartian level on that we find a really spectacular increase in brain size. In less than two million years, brains become three times larger. This is one reason to suggest that language existed as early as in Dartian society.

In this way, anthropologists seek to clothe the petrified bones in flesh and skin and to listen to the voices of those long since dead. It may be nothing more than a thin web of theories that will be ripped apart by the next critic. Still, there are some primary data left, and foremost among them is the enigmatic face of *Ramapithecus* himself.

PART 3

MAN
IN THE ICE AGE

THE STAGE

There have been many ice ages in the long history of the earth. We know little about the earliest ones, but it is well known that there was a great ice age preceding the Cambrian period, some 600 million years ago, and another in the Carboniferous and Permian periods about 300 million years ago. There are unmistakable glacial drift deposits and ice-polished rock surfaces buried among the deposits from those times.

Then, a million years ago or so, still another ice age commenced, and it is still going on. We may refer to it as the Pleistocene Ice Age. It can be seen as the logical outcome of a long-term cooling trend that goes back to the early Tertiary, some 50–60 million years ago, when tropical forests clothed the coast of the North Sea. The cooling was slow and irregular; there are ups and downs, but the downs predominate. In the Miocene, palms grew north of the Alps for the last time; in the Pliocene, crocodiles vanished from central Europe.

Antarctica and Greenland may have been glaciated as early as the Miocene and Pliocene, so that in a way the ice age may be said to have started well before the Pleistocene. But it was only in the middle and late Pleistocene, beginning about a million years ago, that great inland ice sheets formed in Eurasia and North America.

By far the greatest continental ice sheet was that of North America, which at its maximum covered about six million square miles and extended from northern Canada to about the 45th and even the 40th parallel. In Europe, the Scandinavian land ice spread to Britain and central Germany, while smaller ice sheets developed on the Alps, Pyrenees, and other mountain chains. Ice fields also formed on the Ural mountains, in eastern Siberia, in southern South America and Tasmania, and in various highlands at lower latitudes.

But the ice age was not a single cold period. On the contrary, there was a continuous alternation between cold and warm. At least four great glaciations occurred during the last million years or so of earth history. Mementos of their existence are glacial drift deposits, end moraines marking the edge of the ice, thick loess strata formed by windblown dust arising from frost weathering, and so on. As water was withdrawn from the oceans to form the vast ice sheets up to two miles in thickness, the sea level went down as much as four hundred feet, exposing a wide expanse of the continental shelf. Eastern Siberia was joined to Alaska by a land bridge, and other land communications were formed, for example between the Sunda Islands (Sumatra, Java, Borneo) and Asia, or between New Guinea and Australia.

Between glaciations the climate was as mild as at the present day, or warmer—these phases are termed interglacials. But even during glaciations there were ups and downs—the milder spells are called interstadials.

Around the continental ice sheets, great areas were in the grip of an intensely cold climate and formed a bleak tundra landscape. Gradually, animals suited to a life under these circumstances evolved. The elephant stock gave rise to the mammoths, the shaggy proboscideans of the north, with their gigantic ivory tusks that could be used to sweep away the snow and so reach the grass beneath. Then there were the wooly rhinos, similarly protected from the cold by their thick hair and such surviving Arctic animals as reindeer, musk oxen, lemmings, Arctic foxes, and polar bears. In interstadial times, when the climate became slightly milder, these animals retreated northward, and central Europe was invaded by a boreal type flora and fauna with pine forests, giant deer, moose, bison, wolverine, and northern lynx. But in the even warmer intervals that we term interglacials, the ice sheets melted away completely, and the Arctic fauna found refuge in the far north. Instead, temperate areas became populated by a great fauna including the straight-tusked elephant, Merck's rhino, red deer, aurochs, hippopotamus, and many other animals.

Farther south, tropical areas were less affected by the climatic changes, which were primarily changes in the amount of rainfall. The animal world of these areas consisted of forms related or ancestral to the present day animals, as well as extinct creatures like the sivatheres and the hoe-tuskers.

But while the temperatures, the climatic belts, and the sea level shifted back and forth, there were also steady one-way trends. While the sea level went down during glaciations and up during interglacials, it also tended to sink regularly throughout the Pleistocene, so that each new cycle was enacted on a level lower than the preceding one. The animals and plants evolved and changed: The mammoth of the early glaciations was not the same as the form that lived in the late Pleistocene. This is true for the other animals, too, and for man. His evolution was indeed spectacular: From the Dartians of the early Pleistocene, it led on through *Homo erectus* of the middle Pleistocene, to the Neanderthalers and modern men of the late Pleistocene.

And so each glaciation and interglacial takes on an individuality of its own. It has its own flora and fauna, its own type of early man, and its own group of stone age cultures. In Europe, for instance, where the sequence is especially well known, the history is as follows.

In the early glaciation—the Günzian—the animals are still of early Pleistocene type. There are no certain traces of man in Europe, but in Africa and Asia robust Dartians and *Homo erectus* are in evidence. In the early interglacial—the Cromerian—man enters Europe (this is the so-called Heidelberg man). About half a million years ago, much of Europe was covered by the ice of the great Elsterian glaciation, but in an interstadial of that cold phase we again find evidence of man's presence, this time in Hungary.

Next comes the Holsteinian or "Great" interglacial, the date of Swanscombe man in England and Steinheim man in Germany (about a quarter million years ago). The fauna has now been rejuvenated since the early Pleistocene, and new species predominate. A big fallow deer is especially characteristic of England, and the aurochs invade Europe.

Then the ice sheets of the Saalian glaciation spread over areas greater than ever before or since. Yet there is still evidence of human occupation, at least in the milder, interstadial phases, for example at Montmaurin in France.

The last interglacial, the Eemian, begins some 100,000 years ago. In that transient phase of warmth, hippopotami wallow in English rivers, and early Neanderthalers make their home in Eurasia and North Africa. Then, about 70,000 years ago, the last of Weichselian glaciation got under way. Oak, hornbeam, and beech vanish from central Europe, and pine forests spread over the land, but they, too, give way to the creeping dwarf birch, and later on to the tundra. Finally the mighty ice sheet stands two miles thick in northern Europe and reaches down to southern England and Berlin. The woolly mammoth and woolly rhino are seen all over the continent, and the Neanderthalers hunt them for food.

There is an interstadial some 30,000–40,000 years ago. During that phase, modern man enters Europe, and soon the Neanderthaler is gone. There follows a final phase with extreme cold, then the ice sheets melt away, and temperate-type plants and animals press into Europe from the southeast. This is the beginning of the Flandrian interglacial, which started some 10,000–12,000 years ago.

That Flandrian interglacial has, up till now, witnessed the change from the Paleolithic stone age to modern civilization. We still live in the Flandrian, but its warmest part, or optimum, has passed, as early as several thousand years ago. In another few thousand years, the next glaciation will be upon us, unless man's influence changes the climate radically. And the alternation between cold and warm should continue for many million years to come, if the Pleistocene Ice Age follows the pattern of earlier ice ages as we know them.

ENTER HOMO

The name *Pithecanthropus* (ape man) was first coined by E. Haeckel, in Darwin's time, for a hypothetical form of transition between apes and men. Later on it was applied by E. Dubois to the early fossil human type found by him in Java in 1891–92. Now we know that nothing like Haeckel's "missing link" ever existed, and Dubois's fossil man is too human to be a genus of its own. Exit *Pithecanthropus;* enter *Homo.*

Dubois's discoveries were made at Trinil, a village in central Java by the Solo River, and were the culmination of a long campaign that had started in 1889 in Sumatra, then moved to eastern Java, and finally to Trinil. The crucial finds were a skullcap, discovered in 1891, and a thighbone excavated the following year. Dubois returned to the Netherlands with his booty, which was later found to contain four additional thighbones of this hominid, as well as an immense harvest of animal bones. These are still being studied by specialists and are yielding new information about the animal life of the Sunda Islands more than half a million years ago.

Dubois estimated the capacity of the fossil braincase at 800–1,000 c.c., which is approximately midway between the values for gorillas and modern men; thus he thought the genus name *Pithecanthropus* appropriate. Haeckel had thought that his "missing link" would have been unable to speak, so he gave it the trivial name *alalus,* or speechless. But Dubois noted that the thighbone was exactly like that of modern man, so that his *"Pithecan-*

106

thropus" must have been fully erect, rather than the peculiar stooping creature envisioned by Haeckel. So Dubois coined, instead, the trivial name *erectus*.

Nowadays it has become clear that this creature, in spite of its small brain, was related very closely to ourselves, and so a separate genus cannot be justified. This means that he must now be called *Homo erectus* although, of course, the name is almost meaningless from an etymological point of view: there is no species of *Homo* that is not erect, and *Homo erectus* was not the first hominid to stand erect.

At home in Leiden, Dubois encountered the usual scepticism when he exhibited his trophies. Many students accepted Dubois' views, but some of the most influential anthropologists, like R. Virchow, tried hard to repudiate them: to him the skull was that of an ape, and the thighbone that of a normal man. Embittered, Dubois locked away his priceless specimens for more than thirty years. It was not until 1927 that his colleagues were permitted to see them again.

Further collecting was done in the thirties by G.H.R. von Koenigswald, who discovered excellent additional evidence of *Homo erectus*, both in the Trinil strata and in the even older Djetis deposits. According to a potassium-argon date of the so-called tektites—a giant swarm of meteorites that hit Australasia in Trinil times—the age of the Trinil specimens is some 600,000 years, while the Djetis form is even older, perhaps upwards of a million years.

In the Djetis, Koenigswald also found sparse remains of a very large-jawed hominid which was given the name *Meganthropus*, but is now usually thought to be a Dartian. After the war, additional specimens of both *Homo erectus* and the Dartian were found.

In the early twenties, J.G. Andersson found pieces of worked quartzite at another important *Homo erectus* site: Choukoutien near Peking. The worked quartzite suggested to him that there might be fossils of early man awaiting discovery. A molar tooth was found here by B. Bohlin in 1927, and in the following years many skulls, jaws, and other bones were excavated at Choukoutien under the leadership of W.C. Pei. In the end, the material collected represented more than forty individuals. Unfortunately, all of it was lost in the war; only casts are now available. In recent years additional specimens have been collected at Choukoutien.

The age of Peking man has long been disputed, but is now usually set as the second (Holsteinian) interglacial, or about a quarter million years. Clearly, this material is much younger than that of Java man, but there is another find from China that may be as old as the Javanese fossils—a skull from Lantian in the province of Shensi.

Homo erectus has also been found in Africa. Three jaws and a skull fragment of a form closely related to Peking man were found by C. Arambourg and R. Hoffstetter at Ternifine near Mascara, Algeria. The age may be about half a million years. Some other sites in North Africa have yielded more fragmentary specimens.

At Olduvai in Tanzania, specimens that seem to be *Homo erectus* have been unearthed at different levels ranging in age from half a million up to about one million years. The earliest are at least as old as the Djetis men of Java, while the youngest would be contemporary with Ternifine man. Finally, some jaws and skull fragments from Swartkrans in the Transvaal—the cave made famous by the large number of robust Dartian remains found there—also appear to belong to *Homo erectus.*

Thus we see that the various forms referred to as *Homo erectus* range in age from about one million to a quarter million years and in space from China to southern Africa. Within this great area, and during those long millennia, much evolution was taking place. The size of the braincase, for instance, increased from about 800–900 c.c. in the early forms to 1000–1100 in the later ones. The change is reflected in the skull profile. The forehead is quite flat in the earliest forms, while the later *erectus* have a vaulted forehead.

On the other hand, there are also many characteristics that remain much the same, such as the powerful eyebrow ridges, a constriction of the skull just behind the eyes, the bun-shaped cranium with well developed nuchal ridges, and a remarkable thickness of the skull roof which contrasts with the rather thin-skulled gracile Dartians. The face is a protruding one with big jaws, although much smaller than in Dartians, and the cheek teeth, too, are smaller. The *Homo erectus* people were about five feet tall—a little more in the men, a little less in the women—and thus were decidedly taller than the pygmy-sized gracile Dartians and were also fully erect in posture.

A rich assortment of stone tools testifies to the skill of *Homo erectus* in stonework, and innumerable remains of deer and other game at Choukoutien show his prowess as a hunter. There are also fossil hearths at Choukoutien, showing that he learned to use fire. There is no evidence of deliberate burial, but the skulls from Choukoutien show that the base was broken up and the brain extracted before the heads were brought into the cave. The brain was evidently eaten: this is the earliest evidence of anything that could possibly be a religious ritual. It may suggest that *Homo erectus* was concerned with the problem of life and death.

Although a great deal is known about the later form of *Homo erectus* that inhabited the rock shelter at Choukoutien, there is less information about the habits of earlier members of the species. At Olduvai, early *Homo erectus* is associated with a stone age culture of the type named for Abbeville on the Somme. The Abbevillian tool kit is characterized by the presence of a so-called hand axe, a pear-shaped or oval tool produced by rough flaking all round, with rather irregular, indented margins. The hand axe is a general-purpose tool which had many uses. Later *Homo erectus*, both at Olduvai and Ternifine, produced a more sophisticated type of hand axe, the Acheulian variant, with straighter and neater margins. There are many other tools in the Acheulian culture—cleavers, scrapers, knife-like flakes, and round stones. Acheulian cultures are known from a wide area, from the Cape in the south to Britain in the north, but not all the Acheulians were *Homo erectus*.

NEANDERTHALERS

The scenic Neander Valley near Düsseldorf in western Germany was named for the clergyman and hymn writer, Joachim Neander. The psalmist would have been astonished to learn what the name Neanderthal would mean to later generations. The sides of the valley are formed by limestone bluffs containing many caves, and in 1856 a quarryman discovered a number of fossil bones in one of the caves. The bones were sent to J.C. Fuhlrott, a high school teacher in nearby Elberfeld.

One can almost imagine Fuhlrott's surprise upon seeing those fossils. The skullcap was long and low with immense eyebrow ridges, and the thighbone was heavy and peculiarly crooked. Nothing like this had previously been known to science, although it was later found that a similar skull had been unearthed in a Gibraltar cave as early as 1848. But nobody had realized the importance of the Gibraltar find, whereas Fuhlrott was quite clear about the significance of the man from Neanderthal. And so 1856 became the year in which paleoanthropological science was born—three years before the appearance in print of Darwin's *Origin of Species*.

Fuhlrott found a collaborator in H. Schaafhausen, an evolutionist and anatomy professor in Bonn who saw the Neanderthaler as an extinct, primitive type of man. The find was presented at a congress in Kassel in 1857, but, as usual, the discovery and the theories were too novel and foreboding to be accepted immediately. Instead, the Neanderthaler was explained away as a microcephalic idiot, a Cossack from the war of 1814, or an ancient Dutchman or Celt. Virchow regarded him as a pathological specimen, crippled by rickets and other maladies, beaten over the head to receive the unnatural swellings over the eyes, and finally suffering from gout in his old age.

Fuhlrott and Schaafhausen remained undaunted, and the victim of all these imaginary misfortunes, the Neanderthaler himself, soon became acceptable in anthropological circles. Although Virchow's authority dominated in Germany, C. Lyell from England visited the site with Fuhlrott and found evidence of the great antiquity of the find, and W. King described the fossil man as a distinct species, *Homo neanderthalensis.*

Then the Gibraltar skull was remembered and was duly presented at a congress in 1864. Two years later E. Dupont found a jaw fragment of a Neanderthaler in a cave at La Naulette near Dinant, Belgium. Other finds came from Bohemia, and in 1882 K.J. Maška discovered the jaw of a child in the Šipka cave in Moravia. Two Neanderthal skeletons were excavated by J. Fraipont and M. Lohest in definitely Pleistocene deposits in the cave at Spy d'Orneau near Namur. Since then many more discoveries have been made.

Gradually it could be shown that the Neanderthalers were comparatively late in time. Most date from the last, or Weichselian, glaciation, or more precisely, its early part up to about 35,000 years ago when this type of man apparently vanished. A few date back to the preceding warm phase, the Eemian interglacial.

But the Neanderthalers have antecedents in Europe, and it might even be said that there was a special European, or perhaps more properly northern Eurasian, line of evolution that got onto a somewhat different tack than that of widespread *Homo erectus* of the south and east.

The earliest known member of this line is Heidelberg man, whose jaw was found in 1907 in the Grafenrain sand pit at Mauer near Heidelberg. The site had long been studied by O. Schoetensack, who alerted the owner and workmen to the possibility of finding early human remains among the animal bones that were constantly being unearthed there. So nobody was surprised when the jaw came to light, and a notary public was called in to testify to the circumstances.

And its age? Fortunately there is no difficulty about that. The great mammalian fauna from Mauer, as well as the stratigraphy of the site, place it firmly in the

Cromerian interglacial, meaning about half a million years ago—well within the time span of *Homo erectus* in Asia and Africa.

But the jaw differs in various ways from that of the typical *Homo erectus*. The teeth, for instance, are relatively small although the jaw itself is quite massive, with a very low, broad, and shallowly notched ascending branch. Such differences from *Homo erectus*, as well as Neanderthal-like characteristics, have been pointed out by many students, most recently by F.C. Howell.

Further proof of special traits in the Europeans of the middle Pleistocene emerged in the sixties when human remains were unearthed at Verteszöllös in Hungary from deposits that are definitely datable as Elster interstadial—a slightly milder phase within the great glaciation that succeeded the Cromerian interglacial. These fossils probably are about 400,000 years old. In the same strata, lots of burned bones were found. These bones represent the earliest definite evidence known to date of the use of fire.

The most important specimen is the back part of a skull. It resembles *Homo erectus* in the development of big ridges for the neck muscles, indicating that this human must have been just as bull-necked as the contemporary Asiatic and African forms. But the angle formed by the roof and floor of the skull is much wider than in *Homo erectus*, indicating that the volume of the brain case was considerably larger. In fact, A. Thomas, the Hungarian anthropologist, estimates it at about 1400 c.c., which is well above the range of *Homo erectus* and large even by present-day standards. So the deductions based on the Mauer jaw turned out to be right: these ancient Central European peoples were evolving in their own direction.

At the end of the Elsterian glaciation, conditions again became warmer, and we enter the Holsteinian interglacial, when the Acheulian hand-axe culture flourished in Europe. There are also flake-tool cultures of the so-called Clactonian type. In the thirties, the discoveries of Steinheim man by F. Berckhemer in Germany, and Swanscombe man by T. Marston in England, identified the creators of the Acheulian hand-axe culture. The age of these fossils is about 250,000 years. Both are associated with Acheulian hand axes much like those produced by *Homo erectus* in Africa, but the skulls themselves are very different from *Homo erectus*.

The Steinheim skull is well preserved except for some damage to the face and the loss of the lower jaw. Here, too, the base had been broken up, showing that the brain had been taken out. It was found in river gravels at Steinheim near Marbach in Württemberg, at the same level as the rich mammalian fauna of Holsteinian date.

The skull is a rather small, gracile head that probably belonged to a woman. The volume of the braincase is relatively small, some 1150 c.c., but the head is rounded with a well-developed forehead. There is no trace of the constriction of the skull behind the eyes seen in *Homo erectus*, and although the eyebrow ridges are well developed, they are much smaller than in contemporary Peking man. The back of the head is also rounded and lacks the strong crests. Many details are fairly similar to those of modern man: for example, the shape of the cheekbone and that of the mastoid process just behind the ear as well as the high skull vault. Other details may point to Neanderthal.

The Swanscombe skull, found in river gravels of the Thames' 100-foot terrace and positively dated as Holsteinian in age, belonged to a somewhat larger individual and so had a larger braincase—it has been estimated at about 1300 c.c. The variation found in the Vertesszöllös, Swanscombe, and Steinheim skulls—1400 c.c., 1300 c.c., and 1150 c.c. respectively—would be a normal one in a population of this type, with the Swanscombe value as an approximate average.

Unfortunately, the Swanscombe skull is far from complete. Originally, two bones were found, the left parietal and the occipital; the right parietal of the same skull was discovered after the war by J. Wymer. The resemblance to Steinheim man is striking, although the skull bones are somewhat thicker (this was presumably a male). As a detail, it may be noted that the ramification pattern of the meningeal artery, which may be studied from the impressions on the inside of the skull bones, is very advanced in comparison with that of *Homo erectus*. This artery supplies the brain with blood, and its complexity reflects that of the brain. The pattern seen in Swanscombe man is of a type encountered even in some modern men.

According to these finds, then, Holsteinian people in Europe had lost some of the ruggedness that characterized the head of Elsterian man, but continued to be relatively large-brained. There is little difficulty in deriving modern man from something like Steinheim-Swanscombe. But they may also be ancestral to Neanderthal man.

There is still quite a step in time from the 200,000—250,000 years of the Holsteinians up to the early true Neanderthalers of the Eemian, some 100,000 years ago. It is bridged by a find from the cave of Montmaurin near Saint-Gaudens in France, which dates either from the end of the Holsteinian or from an interstadial in the Saalian glaciation. Its age may be roughly set at 150,000–200,000 years. As H.V. Vallois has noted, the jaw would fit the Steinheim skull very well. At the same time it has some characteristics—such as the robust build and small teeth—that are reminiscent of the even older Mauer jaw.

Although physically quite unlike *Homo erectus,* these Europeans had the same type of culture, which again shows that the concepts of race and culture should never be confused.

The Acheulians of Swanscombe made their camps along the banks of the Thames, and finds of charcoal in the area are thought to have come from their hearths. At Hoxne (another site of the same age), there is evidence suggesting that the forest was deliberately fired. Pollen diagrams show a sudden expansion of grasslands at the same level as lumps of charcoal and Acheulian implements.

During most of the later Pleistocene Ice Age—the Eemian interglacial and the early part of the Weichselian glaciation, or roughly from 100,000 to 35,000 years B.P. (before the present)—Europe and neighboring areas were ruled by Neanderthalers. The "classical" or extreme Neanderthal man, exemplified by Fuhlrott's original find, turned out to be the typical European of the earlier Weichselian glaciation. There are many finds, mostly from cave deposits in Europe, for example, the complete skeleton from La Chapelle-aux-Saints in southwestern France. This was an old, almost toothless man, buried in sleeping position in a shallow grave in the cave floor with a bison leg to sustain him during his journey in the unknown.

The physical characteristics of these people were

peculiar. They were short—men on an average were slightly over and women slightly under five feet tall—but they were exceedingly robust and powerful. The size of the head is remarkable. It is enormously long, much longer than the head of a normal modern man, and the skull looks low and narrow because of its length, although in fact it is just as high and broad as that of most modern men. From this it may be concluded that the braincase of the Neanderthalers was exceptionally large, and this is true even though the skull bone is very thick. The brain averages some 1450 c.c., which is somewhat more than that of modern man. For his size, at any rate, the Neanderthaler had a bigger brain than we do.

Sheer size, however, does not necessarily indicate superior intelligence, for the structure of the brain is important too. Much larger than that of man, the elephant brain is one of the largest brains in existence. Yet, clearly, many other animals are more intelligent than elephants.

In the Neanderthaler, the excess size falls mainly in the back part of the brain, whereas the most important brain centers for intelligence are in the fore part. The latter may have been somewhat more primitive than in modern man, but the true meaning of such differences is very difficult to assess. The level of intelligence varies greatly in modern man, and the same was probably true for Neanderthal man. The best might have been on a par with geniuses in modern times, given the same advantages. The Neanderthaler's world, however, was not one in which geniuses would have flourished.

Like the braincase, the face is big, with well-developed eye sockets and nasal cavities. The eye sockets are almost circular in shape, and the upper jaw seems inflated. The cheekbones and lower jaws are powerful and, like the temporal lines, indicate that the chewing muscles were very strong.

Because of the heaviness of the face, the neck muscles had to be strong in order to balance the head; thus we find familiar ridges in the back corner of the skull that indicate a bull neck. The hands and feet are big, but the arms and legs are rather short. Most long bones are slightly crooked, and the two bones of the forearm are more widely spaced than in ourselves. Obviously the Neanderthaler had great physical strength. His was a compact, robust body build, probably excellently suited for bursting forth through the dense scrub forest.

The special culture of the Neanderthaler is the Mousterian, a flake culture (i.e. a culture in which many implements were made from stone flakes) typical of a hunting people, and the animal bones left in the caves where he lived also testify to his hunting habits. At the same time, the very robust dentition may suggest that, like the robust Dartians, the Neanderthalers were dependent on vegetables for part of the year.

The Neanderthalers knew how to use fire—fossil hearths and burned bones are common among the remains. They were also cannibals at times. At Krapina in Yugoslavia, many Neanderthal bones have been found, some of them split open and charred. In a cave in Monte Circio near Rome, a Neanderthal skull was found with a broken-up base, suggesting that it may have been used as a ritual drinking vessel.

Neanderthalian table manners are revealed by a microscopic study of front teeth, discussed by F.E. Koby. The enamel is defaced by innumerable parallel scratches, which were produced when the food was stuffed into the mouth and the part that didn't fit in was cut off with a flint knife. The knife would touch the teeth and leave a scratch. This phenomenon is also known among the Eskimos and other peoples. The direction of the scratches shows that the Neanderthalers were right-handed. A vivid picture is suddenly called forth, showing these savage cave-dwellers congregating in a hungry circle around a recent kill. There are cries and talk, the booming and flickering light of the fire, the sizzling of burning fat, and the smell of roasting, smoke, and drying skins.

All this deals mainly with the glacial Neanderthalers who lived in the earlier half of the last glaciation. Their predecessors in the Eemian interglacial are in some ways less extreme and somewhat reminiscent of the even earlier Steinheim-Swanscombe group.

In the area around Weimar in eastern Germany, there are thick deposits of a calcareous tufa called travertine, which was laid down by springs in interglacial times. Fossil bones are very common in the travertines, and among them are a number of human skull and jaw remains. Animals include red deer, roe deer, fallow deer, lynx, wild cat, straight-tusked elephant, and other typical members of the interglacial forest fauna of the Eemian.

Other finds come from the Eemian desposits at Saccopastore in Italy. Like the German fossils, Saccopastore man is less extreme than the later Neanderthalers. The skulls are more rounded, eyebrow ridges less protruding, and faces somewhat less inflated. A most remarkable pair of specimens comes from Eemian deposits in the cave of Fontéchevade in France. Although relatively flat-headed, Fontéchevade man has a steep and modern-looking forehead, and one of the two fragmentary skulls shows complete absence of the eyebrow ridge so typical of Neanderthal man. These skulls are obviously quite close to modern man.

All of these specimens of Eemian date are 80,000–100,000 years old and so indicate a carry-over from the even older Holsteinians of Steinheim and Swanscombe. There seems to be a line of evolution leading by gradual changes from early immigrants of the Heidelberg type to the characteristic extreme Neanderthalers of the last glaciation. But at that point the line comes to an end. After about 35,000 B.P. the Neanderthalers are gone. There has been much speculation about their fate, but more about that later.

Remains of peoples closely allied to Neanderthal man are known in North Africa, the Levant, and eastwards through Iraq and Uzbekistan to Mongolia and China. The Chinese Neanderthaler is represented by a skull from Mapa in Kwantung and may be somewhat older than the Eemian, though not as old as Peking man. The Mongolian find, from Ordos, is just a tooth. In Uzbekistan, the skeleton of a buried child was dug up in the cave of Teshik Tash. The big Shanidar cave of northern Iraq has yielded several skeletons of Neanderthalers, adult as well as children. Some were found lying where they were crushed under rocks that had fallen from the roof.

In the caves of Mount Carmel, Shukbah, and Zuttiyeh in Israel, and at Ksar Akil in Lebanon, other Neanderthal-like men have been unearthed. These eastern Neanderthalers, though contemporary with those in Europe, are slightly less extreme on various accounts, such as the development of the eyebrow ridges—in the Europeans these tend to form a single bar across the whole face, in the eastern ones the ridges over the two eyes are separate.

MODERN MAN

The appearance of modern man in Europe about 35,000 years ago is very sudden: There is no known transition from the Neanderthalers to those essentially modern-looking people who have been called Cro-Magnon men. Furthermore, these new Europeans are definitely not some kind of "generalized" *Homo sapiens*, but clearly belong to the Caucasoid, or white race. At this early date, then, man had already split up into distinct races. In the same way, the earliest modern men in China are recognizable as Mongoloid, those of Australasia are related to the living Australian aborigines, and early South Africans seem to be allied to the Bushmen. Where did they all come from?

Homo sapiens is easily distinguished from typical Neanderthal men. The eyebrow ridges are weak or absent, the forehead is steeper, the back of the head more rounded, the eye sockets angular, the face less protruding, the teeth smaller, and the chin more prominent. The braincase is on an average somewhat smaller than in the Neanderthalers, but it varies in extreme cases from about 1000 c.c. to twice that volume. The skull bones are thinner and the face more delicate than in the Neanderthalers. The limb bones, too, are more delicate and the chest flatter and less barrel-like. Many other anatomical details could be added.

In 1868 L. Lartet discovered the first evidence of *Homo sapiens* from the Pleistocene at a site known as Cro-Magnon near Les Eyzies in Dordogne, France. The site was a rock shelter, and the remains of at least five persons were found. The best-preserved skeleton is that of a man approximately fifty years of age. The age of the finds can be estimated at about 25,000 years. These people are characterized by their great height, long heads, wide faces, and low, broad eye sockets.

Several other finds from the later part of the Weichselian glaciation in Europe show resemblances to the men from Cro-Magnon. This is true of such early finds as the "Red Lady of Paviland" found in South Wales by Buckland in 1823, the remains from Engis and Engihoul near Liège, Belgium, found by P.C. Schmerling in 1829, and the partial skeleton from La Madeleine, across the river Vézère from Cro-Magnon, discovered by E. Lartet (the father of L. Lartet) in 1864. Later finds include Combe-Capelle in France, which may be upwards of 35,000 years old and probably the oldest find of *Homo sapiens* in Europe, the Grimaldi caves on the Riviera, Predmost and Brno in Moravia, and many others, not only in Europe but also in North Africa. Closely similar men persisted to modern times in the Guanches of the Canary Islands.

Some of the Weichselian Europeans, however, differ in some respects from typical Cro-Magnon men. At the Grotte des Enfants, one of the Grimaldi caves, the skeletons of a middle-aged woman and a sixteen-year-old boy were found. They had been buried together, and have somewhat Negroid features. Whether this really means they are related to Negroes is doubtful, however, and the same is true for the presumed Eskimo affinities of a skeleton from Chancelade in Dordogne. The resemblance in the latter is not very close and probably is due to convergence, i.e., adaptation to a similar environment.

The physical change from Neanderthalers to modern *Homo sapiens* is correlated with a cultural change—from the Mousterian tradition to the late Paleolithic blade cultures—and this change took place approximately 35,000 years ago. There is a striking increase in the material richness and beauty of the cultural remains, enhanced by the appearance of cave art.

In the Middle East the cultural situation is rather different. There are finds of both Neanderthalers and *Homo sapiens,* but both are associated with the same type of culture, a middle Paleolithic tradition lacking the typical late Paleolithic blade implements. The situation is, in a way, reminiscent of the much earlier Acheulian culture, which is associated with *Homo erectus* in Africa, but with pre-Neanderthalers in Europe.

Besides the Neanderthal men from Mount Carmel, no less that ten *Homo sapiens* skeletons (mostly incomplete) have been unearthed in one of the caves, the Skhul cave. There were five men, two women, and three children, all of whom had been deliberately buried. One of the children, a four-year-old, reveals an odd incident. Some time after the burial, somebody started digging a hole and happened upon the corpse of the child, causing some damage to the skull. At that point, however, the unknown digger stopped, perhaps in dread of disturbing the peace of the dead, and filled the pit up again.

The men from Skhul Cave existed as early as the earliest *Homo sapiens* in Europe—dating from the great interstadial of the Weichselian glaciation, about 35,000 years ago—and they resembled Cro-Magnons in many ways, including their fairly great height. But there are also characteristics somewhat reminiscent of Neanderthalers, especially the powerful eyebrow ridges. Yet, when Skhul man existed, the Middle East Neanderthalers were already gone—those from Mount Carmel, for instance, are several thousand years older. Possibly these resemblances may represent a lingering Neanderthal heritage due to crossing between the first invading *Homo sapiens* and local Neanderthalers. However, Professor W.W. Howells informs me that his restudy of the material does not support the suggested Neanderthal affinities of Skhul man.

Indeed, the retaining of the ancient cultural tradition may point to some kind of connection between the Neanderthalers and their successors in the Middle East, for this ancient culture continued to flourish for thousands of years after the demise of its original authors. Finally it was superseded by a late Paleolithic blade culture at the end of the interstadial.

In East Asia the ancient *Homo erectus* and the Mapa "Neanderthaler" were also succeeded by *Homo sapiens*, this time in the guise of early Mongoloid peoples. Typical remains are those from the Upper Cave at Choukoutien, very close to the site of Peking man but much more recent in age. The three skulls found here were originally classified as representing three different types of men—Mongols, Melanesians, and Eskimos—which would indeed be an amazing coincidence. More recent studies suggest that all of the skulls belonged to a single Mongoloid tribe.

In Southeastern Asia, or at any rate in the Sunda Islands, very primitive peoples survived well into the late Pleistocene. This is Solo man of Java, occasionally referred to as a "tropical Neanderthaler" but perhaps more properly regarded as a late survivor of *Homo erectus*. The fossil remains, which were excavated in the early thirties by W.F.F. Oppenoorth and others, consists of eleven skull caps and two shinbones. One of the skulls still shows the fracture caused by the killing blow. The faces, jaws, and skull bases are missing, showing that these are the remains of a cannibal feast. The skulls are very long and low with enormous eyebrow ridges and a deep constriction behind the eyes. Estimated braincase volumes vary between 1150 c.c. and 1300 c.c., indicating an advance from the Peking man stage. The brain is smaller than in the Neanderthalers, however. Despite these crude skulls, Solo man had quite a modern shin, indicating that he was somewhat taller than the Neanderthaler.

In this area, too, the change from a primitive being to an advanced one is very abrupt. Solo man vanished, and in his place we begin to find a number of quite modern skulls, the oldest of which is that of a teen-ager from Niah Great Cave in Sarawak, radiocarbon dated at about 40,000 years. This skull is older than any of the European or Middle East *Homo sapiens,* and in fact is the oldest radiometrically dated skull of modern man known to date. It resembles Tasmanian and Australian aborigines and is probably a member of the population that gave rise to the first immigrants to Australia. There are radiocarbon dates showing that Australia was colonized some 30,000–35,000 years ago. Other early finds come from Keilor in Australia and Wadjak in eastern Java (one of Dubois' early discoveries), and they also resemble the Australian group.

While North Africa has a history resembling that of Europe—Neanderthal country in the early part of the Weichselian, Cro-Magnon in the later—Africa south of Sahara seems to have been populated by late *Homo erectus* survivors much like Solo man. The earliest find, in 1921, came from Broken Hill in present-day Zambia and consisted of a skeleton found in a deep narrow cave. Like Solo man, it is of late Pleistocene age and probably no older than 30,000–40,000 years. In other words, Broken Hill man lived at the time when *Homo sapiens* spread far and wide from Europe and North Africa in the west to China and Australia in the east and south.

But Broken Hill man is definitely not a *Homo sapiens*. The skull is very long, the eyebrow ridges enormous, and the face very big with a particularly broad nasal opening. The volume of the braincase is 1300 c.c. An unusual detail is the caries-ravaged dentition. But although his head is so primitive, his body skeleton is modern-looking and slim-boned and completely lacking in specialized Neanderthal characteristics.

A second skull of this type of man was discovered in 1953 at Hopefield, Saldanha Bay, South Africa. This specimen is only a skullcap, but is essentially similar to the Zambian form, showing that men of this type ranged widely over sub-Saharan Africa. Their relationship to Solo man may be fairly close.

These late *Homo erectus* survivors were succeded in southern Africa by men of *Homo sapiens* type. The earliest find in this area seems to be Florisbad man, who may be about 35,000 years old. His skull was found at a warm spring site near Bloemfontein, and shows a mixture of advanced and primitive characteristics. The latter, such as the rather receding forehead and the fairly well-developed eyebrow ridges, may be due to a mixed heritage, as in the case of the Skhul men of the Middle East.

A later invasion may be represented by the skull from Boskop in the Transvaal and various similar remains from other sites. The five-cornered shape of the skull seen from above, and the small, infantile face, resemble that of the living Bushmen, but the Boskop skull is quite a bit larger than the Bushman's. However, a reduction in size since the Pleistocene Ice Age is so common in many mammals all over the world that the pygmy stature of the present-day Bushmen may well have evolved from normal-sized ancestors of the Boskop type.

At this point, we are presented with the outlines of a major problem. Almost everywhere, either Neanderthalers or late *Homo erectus* were replaced more or less suddenly by modern types of *Homo sapiens*. Moreover, these invaders show different racial traits which must have taken some time to evolve. But where did *Homo sapiens* originate?

One theory, formulated most recently by C.S. Coon, suggests that men of modern type evolved locally from Neanderthalers in the north and from men of the Solo and Broken Hill type in the south. A theoretical case may be made for this, but I have to admit I cannot find it convincing. The change is too great, the time too short, and there is nothing like it in previous times or in other evolving lineages. In Europe there would be a change from extreme Neanderthalers to Cro-Magnon men—two types of men that are poles apart in body build—within 5,000–10,000 years while the much smaller change from the Steinheim-Swanscombe type to the extreme Neanderthaler would have lasted 200,000 years. In the south, with the late dates for Solo man and Broken Hill man, the change would be, if anything, even more abrupt. I do not say this is impossible, but it smacks too much of coincidence and involves too much special pleading to be palatable to the evolutionist.

In addition, we do have some evidence of a greater age for typical *Homo sapiens* than the 30,000–40,000 years ago when the great takeover took place. First, the 40,000-year-old boy or girl from Great Niah Cave is definitely as old as some Neanderthalers and southern *Homo erectus,* or older. Then there are the finds from the Kan-jera deposits in Kenya, discovered by Leakey in 1932. They consist of four skulls and a thighbone, all of them quite modern in type except that the skull bone is rather thick. Their age, judging from the associated animal bones, is at least 60,000–70,000 years. It has been ar-gued that their presence in the old deposit occurred by burial, but studies of the chemical composition of the bones and that of animals from the same deposits sug-gest that they were contemporaneous. Recently, very early remains of *Homo sapiens* have been found at Omo in Ethiopia.

But there are even older fossils of the *Homo sapiens* type. These are the two skulls from Fontéchevade in France mentioned in the previous chapter. As H.V. Val-lois has shown, the development of the forehead and eyebrow region in Fontéchevade man is not only com-pletely unlike that in Neanderthal man, but also similar to that in modern man. We can hardly regard them as aberrant Neanderthalers.

Attempts have also been made to dispute the age of Fontéchevade man, but the evidence for an Eemian age is quite conclusive. The two skulls were found in 1947 by experienced excavators under the highly competent leadership of Germaine Henri-Martin. They were in a deposit containing typical Eemian interglacial mammals, and the deposit was sealed in by a thick, undisturbed layer of dripstone which makes a later burial impossible. So we cannot get around the fact that *Homo sapiens,* at least in an early guise, was present in Europe in the Eemian interglacial most probably in its earlier part, about 100,000 years ago.

This is enough to show that the modern type of man has been in existence for a fairly long stretch of time and probably can be derived from something pretty close to Steinheim and Swanscombe men of the Holsteinian. The main question is: where was the original homeland of this stock? We simply do not know at present.

Regardless of its origin, this modern type of *Homo sapiens* began to set out in various directions more than 40,000 years ago. As wandering tribes colonized new areas, they evolved in adaptation to local conditions and so became racially different. As we have seen, most of the great living races of man developed at an early date—the Australoids, the Mongoloids, the Caucasoids, and at least early variants of the Bushmen, or Capoids. The only great race not yet known from early times is the Negroid, but there is no reason to believe them younger than the other races. As W.W. Howells has shown, they probably differentiated in the West African rain-forest area where the fossil documentation at present is incomplete.

THE ALIEN SPECIES

There are two views on human evolution. In one of them, mankind has split up repeatedly into distinct species—that is to say, hominids that did not breed with each other and presumably did not regard the other species as "men" (but perhaps as trolls). Later on, all of the species except one became extinct, and the extinction was usually due to competition, if not out-and-out intergroup struggles.

In the other view, there has never been more than one species of man at a time. Different-looking hominids living at the same time are regarded as individual variants or, at most, different races. Interbreeding never ceases, and when a type of man—like the Solo form or the Neanderthaler—vanishes, this is most probably due to their being swamped by more numerous invaders.

The facts in this case may be of more than token interest, for they have a bearing on our future, as I hope to show in a later chapter.

Those who believe in more than one species at a time point to the great differences between hominids that lived during the same periods, such as robust Dartians and Swartkrans *Homo erectus,* or Neanderthalers and modern *Homo sapiens,* and suggest that this shows they were specifically different. Those who believe in only one species point to human sexual behavior, which leads to interbreeding whenever different stocks meet. No stock of human beings, they say, really succeeds in remaining isolated long enough to become a truly distinct species.

It is striking that—as one of the one-species men points out—the "bone-oriented" scientists generally tend to favor the more-than-one species alternative, while "theory-oriented" students tend to be single-species men (substitute "paleo-anthropologists" and "neo-anthropologists" respectively, and you will see what I mean). Our one-species man goes on to make fun of the supposed importance of "a bump here or a centimeter there" and states flatly that the course of human evolution cannot really be understood by comparative anatomy or by studies of the "miserable scraps of bone" that constitute the fossil material.

From the foregoing, it should be clear that I view such statements as this with complete disgust. Too many sweeping theories have been followed by increasingly desperate attempts to get unruly fossils to toe the line. We can confidently turn the statement quoted above inside out and say that we shall never learn anything about how man evolved except on the basis of the fossil evidence. We need theories, but the theories have to fit the historical facts.

So to decide the question of one species or several, we again have to see what the fossils tell us. Let us start with the Dartian group, which is divided into two basic forms—gracile and robust Dartians. One species or two? Could they simply have been male and female of a single species? In that case we would have an all-female tribe at Sterkfontein, and an all-male one at Swartkrans.

The evidence in hand shows that both Dartian forms are of great antiquity and existed in the same general area for a long time, at Omo and again at Olduvai. It also appears that robust Dartians persist in even later times when higher hominids had already appeared—at Swartkrans, for example.

To this, the one-species man may reply that the different types are really just variants of a single population and that we underestimate the variation possible in a human population. Modern man is indeed extremely variable. It would be possible to find two modern skulls that differ quite as much as, say, those of a robust and a gracile Dartian—yet they are both of the same species, *Homo sapiens*.

But in reality there is good reason to suspect that local Dartian populations were much less variable than the entire range of modern humanity. In the first place, Dartians were certainly very few; I doubt whether as many as one million were ever alive at one time. In a population consisting of a single person, the variation is obviously nil. If there are two persons, the variation may be great or slight, but probably is moderate. By and large, the more you add to the number of individuals, the greater the variation. This becomes obvious when you compare a Dartian population of less than a million with modern humanity numbering several billion.

We must also remember that modern man is distributed into many local populations which for a long time have been adapting to a variety of life conditions—life in the Arctic or the tropical jungles, in the deserts or mountains, islands, lakes, and so on, as well as a multitude of different economies—hunting, fishing, collecting, farming, trading, etc. In contrast, the local Dartian population must have been comparatively uniform, living in a restricted area. In such a situation its variation must have been relatively slight, and we have no right to assume that it even remotely approached the range seen in modern man.

The fact that robust and gracile Dartians existed as separate forms for millions of years indicates that there was little or no interbreeding, or else they would simply have merged into one form. As this is how species are defined, we may say that they were distinct species.

With this the whole one-species theory becomes doubtful. Probably the sexual behavior of early hominids was somewhat different from what it is thought to have been. The mysterious process of object-fixation or imprinting may have a great deal to do with mapping the way of future sexual activity. This fixation is a kind of "instantaneous learning" occurring at the first release of an instinctive drive. Obviously, in a uniracial society there will be little stimulus to regard very different-looking creatures as "men" or mating partners.

How about later men? That *Homo erectus* was a species different from contemporary robust Dartians is clear enough from the evidence at Swartkrans and Olduvai. But were the mid-Pleistocene Europeans of Heidelberg, Vertesszöllös, Swanscombe, and Steinheim a species different from *Homo erectus* of the Choukoutien type? If not, they were certainly on their way to becoming so.

When we come to the late Pleistocene, and to modern *Homo sapiens* versus the lingering *erectus* men of Solo and Broken Hill, it seems really difficult to escape the conclusion that here indeed were two distinct species. The same might possibly be held, too, for the extreme Neanderthalers of the last glaciation in Europe; they vanished without a trace of hybridization, and suggested evidence of hybridization in the Middle East now seems doubtful, also. This interpretation is strengthened by the presence of *Homo sapiens* at Fontéchevade, long before the appearance of the extreme Neanderthalers.

But even if this appears to be the most likely interpretation, I do not wish to be dogmatic about it, and it is not necessary, for the Dartians have already proved the point: mankind can split up into distinct species that do not regard each other as "men."

BRAINS

There are three main characteristics distinguishing men from apes. Of these, the basically human dentition has been with us from the start. The upright posture probably came in Pliocene times, between five and ten million years ago. The large brain, last of all to appear, evolved in the Pleistocene. Many bits and pieces of vital information are still missing, but there is sufficient evidence in hand to map at least some of the outlines of evolution of the brain's size in the Pleistocene. The story can be told as follows.

There was probably some increase in average brain size during the long millions of years of Dartian existence, but it was extremely slow. About two million years ago the average braincase volume was about 500 c.c., and the robust Dartians that survived in mid-Pleistocene times—more than a million years later—had only slightly larger brains.

While this was going on, however, one group of evolving humans exhibited an increased rate of evolution as far as brain size is concerned. This is the stock called *Homo erectus*. When we first meet them about one million years ago, the size of the braincase had already reached about 750 c.c., and it continued to increase at a very regular rate, about 4.6 per cent per 100,000 years. About 400,000 years ago, it had reached the 1000 c.c. level, on an average, and when this line came to an end in the late Pleistocene with the Solo and Broken Hill types, the average had risen to 1100–1200 c.c. It is rather intriguing to see that this increase in the size of the brain progressed at a constant rate within this stock for about one million years.

If we try to derive *Homo erectus* from the Dartians, we come to the conclusion that it probably developed two or three million years ago and that what started it off was apparently an increase in the rate of brain evolution. We may suspect that something happened at the time— something that set up a new situation, changed the conditions of life, and made brain size much more important than it had been previously. Moreover, it was something that was missed by the robust Dartians, who remained small-brained until the end of their time.

What happened? Was it a cultural innovation that put an added premium on intelligence? This seems likely enough, and, in fact, the earliest evidence of stonework known at this time comes from deposits at Lake Rudolf, dated at 2.6 million years ago. This culture is fairly varied and must be the result of a long history of evolution. So perhaps this was the factor that gave natural selection a new impetus.

There is at least one more possibility: the invention of articulated speech. Many authorities have been reluctant to admit that such primitive beings as the Dartians, and even *Homo erectus*, could speak. But the recent experiments of the Gardners with chimpanzees show that apes can learn a sign language of considerable complexity, and there is certainly every reason to believe that the Dartians were more intelligent and above all more language-minded than chimps. So the decisive factor may well have been the emergence of speech. Or speech and stonework may even have evolved together. While the later *Homo erectus* followed their evolutionary course, history was repeated once more. From some early form of the *erectus* stock, a new evolving group branched out represented by Heidelberg-Vertesszöllös man and the Holsteinians to the late Pleistocene Neanderthalers. Here the rate of increase in brain size became even higher: it may be estimated at 7.5 per cent per 100,000 years at minimum, but probably was even more rapid.

The Neanderthal stock must have departed from the *erectus* lineage no earlier than 800,000 years ago and probably somewhat later. It culminated in the glacial Neanderthalers, the large-brained men who lived in northwestern Eurasia during the early part of the last glaciation some 50,000 years ago, and whose capacious braincases measured some 1500 c.c. on an average.

Apparently something happened in their ancestry, too. What could it be? Another cultural invention? The use of fire comes to mind. It is recorded in this lineage earlier than in any other: Vertesszöllös man used fire 400,000 years ago. But we must also remember that *Homo erectus* used fire, too, although at a somewhat later date: Choukoutien man, for example, used fire 250,000–300,000 years ago. This would be early enough to influence the final stages of evolution in the *erectus* stock, but apparently it did not.

We can see how speculations of this kind, fascinating though they may be, tend to lead us into blind alleys. We can make guesses, and it is legitimate to do so, but we do not know for sure. We can only say that, based on the evidence at hand, it seems that the evolution of brain size was suddenly accelerated at least twice during Pleistocene times.

The story, of course, does not end with Neanderthal man; we then have to account for the rise of modern *Homo sapiens*. It is perhaps reasonable to derive him from ancestral, smaller-brained pre-Neanderthalers such as the Holsteinians of Europe. From this group, his evolution to the modern average of 1350 c.c. would have proceeded at about the same rate as in *Homo erectus*, though starting from a higher level, so that his brains would have been on an average about 150 c.c. larger than those of contemporary *Homo erectus*.

Thus skulls became more and more inflated as time passed by. There are other changes too, but they are less regular. In some types of man the skull walls are very thick, as in *Homo erectus* and the Neanderthalers; in others they are thin, as in gracile Dartians and modern men. These differences are probably adaptive, and it has been suggested that a thick skull vault evolved in those types of men for whom the bludgeon was the main weapon. With spears and other stabbing weapons coming into fashion, the protection given by a thick skull became less important, and the lightness and economy of a thin skull was favored.

But the basic factor is still the increase in the size of the brain itself, and at this point we must ask ourselves why the human brain should be so big. It would seem that our brain functions vastly exceed everything that is reasonably needed in the environment where they evolved—that of the Paleolithic hunters and gatherers. Why should we be able to grasp advanced mathematics and philosophy and enjoy intricate poetry and music— things that cannot possibly have played any part in the life of the Paleolithic man?

The answer is really quite simple. Modern man can do these and other complicated things, but only if he has had the benefit of a good education. Natural selection, however, is always working with the material at hand, and the material of Paleolithic times consisted of more or less untutored brains, which were able to use only a small fraction of their total capacities. To produce an increase in the "available" skill of the brain, natural selection had to call forth a vastly greater unrealized capacity, which remains dormant in the savage but can be activated by suitable education.

In a corresponding way, intelligent animals like apes, dogs, and dolphins can be taught extremely difficult chores, thus activating brain resources not normally used in the wild. As with the proverbial iceberg, perhaps nine-tenths of the total ability of the brain may be not apparent. There are even those who think that dolphin brains may turn out to be superior to human ones.

There is an additional factor that might be called the perfectionism of natural selection. We can see how it works in some cases of mimicry, where a harmless or tasty animal deceives hungry enemies by mimicking some poisonous or bad-tasting animal. In some instances the mimicry is far more precise than what would be necessary to achieve the deception. This is because every tiny improvement will give its possessor a small selective advantage, and so will become incorporated into the fabric of deception. Then again, look at the incredibly detailed adaptations to each other as seen in some species of orchids and insects—a topic that fascinated Darwin and has continued to attract students. It is this insistence that a good thing can always be improved upon (if just a little bit) that probably played an important part in the evolution of the human brain. The principle is simple enough. *Homo erectus* was intelligent enough to get along very well in his environment. But *Homo sapiens* was even smarter and so he crowded out *Homo erectus*. Within each population the principle worked in the same way. The average Neanderthaler did well enough, but the brightest Neanderthalers did even better.

In this way orchids, insects, and dolphins, fossil skulls and chipped stones—in the world around us today and in the world of a million years ago—help to explain the emergence of man's third and most important characteristic: his big brain.

PART 4

HERE
AND NOW

SELECTION

So far we have talked of man's past; now it is time to turn to the present and future. This is an inexhaustible topic, and I intend to investigate it only from the point of view of the evolutionist; even so, there is much to be said.

Up until very recently, man—at least in the Western world—has tended to see himself as the lord of creation, smugly reposing in a sphere apart from the world of strife and struggle for existence where other living beings dwell. Belief in this view has now been rudely shaken by a growing general awareness of the population explosion, of our helplessness in trying to get society into shape, and of the ever increasing pollution of our environment.

But even those who realize all this may still think that man is exempt from the laws of nature that govern natural selection and evolution. The theory of natural selection was set forth by Darwin in 1859 and overnight became a hotly debated topic. In our time, however probably few people other than biologists have any interest in it, and even fewer think that it might have some bearing on the situation of modern man.

Perhaps this is so because most of Darwin's contemporaries were mainly impressed by the kind of selection that is represented by differential mortality—the very struggle for existence: survival is the lot of the most intelligent, strong, and skillful; premature death, that of the less well-equipped. To apply this to human beings, however, appeared so abhorrent that some people altogether refused to accept the idea of natural selection and substituted some "nicer" machinery of evolution. It might be the inheritance of acquired characteristics—surely a pleasing thought: by improving yourself, you produce better offspring. But it does not work that way, for acquired characteristics are not inherited: they have to be acquired anew by the next generation.

Another semi-solution was the idea of an inner force of evolution, the *élan vital*, which would automatically carry us forwards to new heights of nobility and spirituality. Unfortunately, there is no evidence whatever for the existence of such a force. Evolution has been shown to come from the interaction between variations in inheritance—resulting from mutation—and natural selection. This is all, and we have to make do with it.

At this point the evolutionary optimist might reason as follows. Natural selection tends, after all, to favor those that are best equipped—the most intelligent, the strongest, the bravest, and perhaps also the most handsome (sexual selection was another one of Darwin's discoveries). Even though the process of selection may bring suffering to the individual, the human stock as a whole will be improved as time goes by, so that our distant descendants will be brighter, more beautiful, and healthier than we are.

Unfortunately, his opponent may now retort, natural selection has been largely put out of action in modern, civilized society. In Paleolithic times the situation was different. Each woman bore a large number of children, but mortality in childhood was very great and, at the same time, strongly selective. As a result, the whole population had a genetic make-up predisposing for intelligence and good health. Intelligence comes in because the brighter parents were more able to care for their children, and the brighter children more able to avoid dangers. The main factor in all this was selective mortality, and the death of so many children was the price paid by society for health and progress.

In the modern world, most children live to become adults and may have children of their own, even if they happen to be nearsighted, deaf, or moronic, or to have some other slight genetic disadvantage. Only those born with really serious defects die young, so it would seem that natural selection is almost powerless and that we can look forward only to gradual degeneration in the future.

As a matter of fact, selection by differential mortality is still very much with us although it has shifted its emphasis to some extent since Paleolithic times, and there is good reason to think that its role is going to increase. But before we go on to explore differential mortality, we must take account of a different kind of selection that has now taken over much of the selective function in human evolution.

This "new" kind of selection is not really new although it has become more important in our time than it was in the past. This is selective fertility. As G. G. Simpson has stated, in a given population, if a random 50 per cent of the parents have 50 per cent of the children, there is no differential birth rate and hence no selection of this kind. But if 50 per cent of the parents have 80 per cent of the children, there might be very strong selection indeed—provided that those 50 per cent with higher fertility also differ in genetic traits from the other 50 per cent.

What kind of people tend to have more children and what kind to have fewer? In most civilized societies the number of children per family has, within recent times, tended to be inverse to the amount of the family's worldly possessions: where lower class homes thronged with offspring, the upper classes had one or two. This, however, seems to be a transient phase in modern civilization, typical perhaps only of the last century or two, and there is some evidence that the difference between upper and lower class child bearing rates is again diminishing. In any case, in order for the difference to be truly selective, there ought to be a genetic difference between the higher and lower classes in society. Whether this is really true is uncertain, and the debate is correspondingly emotional and biased.

Probably much more important, however, is the differential fertility within the population of the world as a whole. Those populations that increase at the highest rate—for instance that of Latin America—will form an ever-increasing fraction of humanity as a whole. While the population increase as such is obviously a serious threat to man's future, the increasing predominance of certain races can hardly be regarded as good or bad so long as we have no objective way to judge the genetic excellence of different races. In any case, it is one of the clearest examples of contemporary natural selection by differential fertility, but it is a case of selection between, and not within, populations.

Such are the main outlines of present-day selection in man. It is now time to go into closer detail, and we turn to the classical Darwinian selection by differential mortality. Far from having disappeared, differential mortality is still a major force in man's evolution and continues in new forms.

THE
BAD GENES

In extreme climates and environments, the old force of selection is still actively molding man's physical characteristics in spite of all our attempts to foil it by introduction of air-conditioning, double windows, and other devices to protect ourselves. To take advantage of them, however, we must resort to a troglodytic existence which is possible only for a small fraction of the population. Most people still have to brave the elements for at least part of the time and have to grin and bear it.

In a hot climate, then, it is still useful to have a lanky build, dark skin, matted hair, and highly efficient sweatglands. It may also be helpful to be small of stature in order to increase the ratio between the surface and the bulk of your body, and if you need body fat as a store of energy, it had better be concentrated in one or two places so as not to interfere with your heat exchange.

Conversely, in a cold climate you may well be fairskinned, so as to get the most of the precious sunrays, and it also helps to be fairly big, or at least stockily built and short-necked, with your body fat spread out evenly to protect you from the cold—especially over the face, which is difficult to cover. A good blood supply to your face and hands will also help keep them warm.

Let the two types of men change places and see how

selection takes over. In the hot climate, heat-stroke may be its favorite weapon, followed by a weakness that renders the victim more susceptible to local diseases, against which he probably does not have an innate protection. In the cold climate it is exposure to low temperatures that saps the resistance to infection.

How this works in practice is well illustrated by the Indians in the South American highlands. Their physique is markedly adapted to their harsh surroundings: their lungs and hearts are very large, and their blood contains a proportion of red corpuscles greater than that of the average human. All of this is necessary to offset the effect of the oxygen-poor air of the highlands. And in spite of massive immigration of other peoples, both Caucasoids and Negroes, the Indian type remains unaffected in the highlands. Immigrants fare badly in the thin air, and even if they survive, stillbirths are common because the growing embryo does not receive sufficient oxygen from its mother, and so it suffocates. Differential mortality is back with us again, filtering away the alien genes that have no business being there.

In this case, then, what we could call the "normal" genes turn out to be "bad" genes—i.e., bad for this special environment. Otherwise, when one speaks of bad genes, diseases caused by genetic defects come to mind. When such diseases are incurable, the problem is handled by differential mortality, but the situation may be more complicated than it looks at first glance.

A typical example of a bad gene is that causing bleeder disease, or hemophilia. This disease, as is known only too well, occurs in the best families (more specifically, in the male offspring of Queen Victoria). The fact that the disease is seen in males only indicates that it is caused by a gene in the sex chromosomes. In females, the two sex chromosomes are similar and are termed X chromosomes, but males have one X chromosome and a somewhat different, so-called Y chromosome. When the female ovum, which contains an X chromosome, is fertilized, the sex of the child is determined by the sex chromosome of the sperm: if it is an X, the child will have XX and will be a girl; if it is a Y, the child will be XY, or a boy.

The hemophilia gene lies in the X chromosome, which means that it is transmitted by the mother to the son; from the father, it can be transmitted only to a daughter. But in a woman, the effect of the deficient gene is covered by that of the normal, corresponding gene in the other X chromosome, and there will be no symptoms. In a man, on the other hand, there is no corresponding gene in the Y chromosome, which simply lacks a number of genes present in the X, and so the disease occurs.

The life of a bleeder is constantly being threatened by the difficulty in stopping bleeding, even from minor wounds, and so, on an average, his life will be shorter and his offspring fewer than in normal males. It would seem then that natural selection is strong against this inherited trait, and it ought to vanish from the population in the course of time. But the story is not quite that simple.

In the first place, mutations from the normal gene to the hemophiliac occur now and then, so that the store of this "bad" gene is constantly being replenished. But there is another even more surprising aspect: when present in a female, the hemophiliac gene actually tends to increase fertility rather than decrease it, and thus tends to be favored by natural selection.

This shows that a gene can be "bad" or "good"

depending upon its environment—the genetic or physical environment. The best known instance of this is sickle-cell anemia, a disease typical of certain malaria-infected areas.

This is a heritable disease, and it occurs only in persons who have received a "sickling gene" from both parents. In persons with this disease, the red blood corpuscles are highly abnormal, being sickle-shaped instead of round, and the disease leads to death at an early age.

No such complications occur in the individuals who have only one sickling gene, having received a normal gene from the other parent. Such persons are completely normal in all respects except one, but that one is the more important: they are resistant to malaria. At the same time, there are many people who have no sickling gene at all, as they have received a normal gene from both parents. These individuals are of course quite normal, and so, they can fall victim to malaria.

There is, then, a selective situation controlled by differential mortality. On the one hand, all of the persons having two sickling genes will die of anemia. On the other hand, many (though not all) persons without any sickling genes will die of malaria. The result of these selective forces is that a certain balance in the frequency of the sickling gene is struck. The sickling gene will remain in a mathematically predictable frequency within the population, and a certain group of people will enjoy immunity to malaria at the price of so many dead of anemia and so many of malaria.

This makes it fully clear that being subjected to natural selection by differential mortality is a most unpleasant situation, to put it mildly. True, this is one of the crassest instances of its kind, but as will be shown later, we have other instances that will appear much more familiar to us.

As for sickle-cell anemia, the situation has changed recently. Treatment of malaria has become more effective, and the mosquitos that transmit the disease have been killed in many areas. As a consequence, the direction of selection has changed. Sickle-cell anemics still die, but malaria patients recover, and fewer people get malaria. Selection now works against the sickling gene and is rapidly reducing its frequency. The population is evolving.

If we were able to cure sickle-cell anemia, on the other hand, the selection against this gene would be reduced or would stop altogether although the "bad" gene would remain present indefinitely. This is, in fact, the result of successful medical cures of various heritable diseases or shortcomings (there is, in fact, new hope of a cure for sickle-cell anemia). Diabetics can live a normal life with the help of insulin. Nearsighted persons use glasses; deaf persons use hearing aids. Selection has been switched off, and the number of "bad" genes is steadily increasing by mutation. Is this good or bad? Which is more important, the happiness of the individual or the health of mankind as a whole?

It may just as well be said that the "bad" genes have ceased to be bad as soon as their effect can be offset by suitable medication. We should remember in this connection that all of mankind suffers from a really "bad" gene: none of us is able to synthesize vitamin C in our bodies. We have to get it in our food, or else we die of scurvy. Most animals do have the capacity to produce this vitamin and so never run the risk of this disease. Man shares his handicap with the other higher primates and with swine and, like these animals, has to eat fresh fruit or some other food rich in vitamin C to keep well. This makes it highly probable, incidentally, that even if early hominids were mainly carnivorous, they included lots of fruits and other vegetables in their diet.

Few people think of this now as a real handicap or as a genetic deficiency, even though it probably led to great suffering for some unfortunate hominids up until quite recent times. But in principle there is little difference between providing lime juice, spectacles, or insulin; in all of these cases a genetic defect is being corrected and a normal life made possible.

Thus the "bad" genes may not really pose a serious threat to humanity. Even more surprising, they may actually be beneficial. Studies of wild animal populations have shown that they contain an amazing number of different genes that, if present in double dose, could lead to death just as surely as the sickling gene. The same is apparently true for man: it has even been estimated that each individual, on an average, carries two lethal genes. What is more, this "genetic load" may be advantageous: we have seen that sickling-gene hybrids are resistant to some diseases and that bleeder-gene hybrid women have higher fertility. The phenomenon is known as "hybrid vigor" and seems, unexpectedly, to be especially pronounced in the case of lethal genes. So then, far from being "bad," they may be essential to the well-being and vigor of the human race.

But the phenomenon of hybrid vigor is also seen in other cases of mixed heritage, and so it may be suggested that an increased mixing of different stocks and races will be beneficial mainly from the genetic point of view. Many laymen probably think of biologists as persons interested in purity of race and the like: this may be because some influential racialists in Nazi Germany had the gall to refer to themselves as biologists. It may come as a surprise to learn that geneticists tend to take the opposite view— they don't believe in the purity of races.

This should be clear to quite a few dog-lovers. How many purebred dogs are nervous wrecks in comparison to the ordinary mutt, a picture of good humor and mental stability?

GENES
AND BEHAVIOR

There is an old German definition of a professor as *ein Herr anderer Meinung,* or a gentleman of a different opinion, and this is particularly true for students of man and his activities. There is a wide range of opinions concerning man's life. At one end is the belief in an inherent fate, a genetically predetermined course of life; at the other, the view of the human mind as a blank slate that is gradually filled by messages from the environment. The truth, as usual, is somewhere in between.

In the animal world as a whole, there is indeed a complete spectrum from entirely programmed to entirely "open" behavior. But students have found that even some of the apparently fixed instincts of quite lowly animals can in some cases be modified by learning, and in a corresponding way we may suspect that some items of seemingly "acquired" behavior may contain some inherited components. Genetically conditioned behavior is, of course, subject to natural selection. The capacity to learn, however, is also determined genetically and is similarly within the realm of selection.

Man's behavior is a mixture of inherited and environment-induced responses and actions. In most physical movements, the programming is quite important. The way we walk, for instance, results from such programming inherent in the constuction of our skeleton, muscles, sense organs, and nervous systems. We say that children "learn" to walk, but we might as well say that they gradually develop their congenital program for the art of walking. The act of walking, however, may be directed towards a purpose that has little or nothing to do with innate behavior—such as walking to the library to borrow a book. Between the two poles, however, fall many other activities of a varying degree of "innateness" and "openness."

Imagine, for instance, that you meet a person who proceeds to pull your hat down over your eyes, kicks you in the shin, and calls you an insulting name. In this situation your suprarenal bodies—two small, ductless glands on top of your kidneys—start pouring into your bloodstream a substance called adrenaline, which has a profound influence on your physiology. It puts off your digestion for the time being, so as not to interfere with more important matters, such as stimulating the beating of your heart and the working of your sweat glands; it also makes your liver suddenly release its energy stores for instant consumption; your hair stands on end, your eyes protrude; you are ready to fight or flee. In other words, you become angry.

What you do with your anger depends on many factors, of which you may at the time be only dimly conscious. If your assailant is very formidable, you may simply take off in flight, but let us assume that he is a little punk you could squash with one hand tied behind your back. If you are a congenital choleric and have not had the benefit of a repressive education, you are likely to become aggressive. If you have a phlegmatic disposition, you are probably going to react more mildly. However, the choleric who has learned self-discipline in the school of life may react in the same way as the phlegmatic, and, conversely, a phlegmatic person whose education has favored militancy may well become aggressive.

This illustrates, if crudely, the effects of inheritance and environment on human behavior. What is inherited is not one trait or the other, but a way to develop in reaction to the environment. The "openness" of man's development is such that most people can be forced into a preconceived mold and function tolerably well in it. You can make almost anybody into a soldier (or farmer, or merchant) provided that the education is efficient and starts early. But some men, because of their genetic constitution, will make better (or at any rate more easily trained) soldiers and will probably be happier in their profession than many others. Fortunately, modern society offers many different occupations, and differently endowed persons are likely to find one that suits their disposition. This must be one of the chief aims of education.

That behavior is influenced in such ways by inheritance can hardly be doubted, but as evolution consists of changes in the genetic constitution of populations, this is the aspect of behavior that we have to examine here. However, it should be noted that this does not mean that educability is denied or forgotten.

In early genetic studies investigators were concerned mainly with clear-cut unit characteristics that were easy to recognize, such as flower color in plants and eye color in fruit flies and man. When it came to human behavioral genetics, the first genes to be studied were again those causing marked abnormalities. For instance, geneticists have concluded that about 80 per cent of the different kinds of imbecility and feeblemindedness are caused by genetic defects, and the percentage is even higher for the two most common mental diseases—manic-depressive psychosis and schizophrenia.

Recent advances in the study of human chromosomes have made it possible to observe changes that affect both behavior and other traits. The normal chromosome number in man, for instance, is 46, but there are individuals who have too few or too many. This results from a disturbance in the reduction division of the forming sex cells, which should have one-half of the chromosome number. If everything goes well, the chromosomes are segregated so that the eggs and sperms carry 23 chromosomes each and, at fertilization, the original number is restored.

But the reduction may miscarry. Both chromosomes of a pair may, for instance, go to one of the daughter cells, giving it an extra chromosome and leaving the other with one less than the correct number. If such a sex cell—an egg or a sperm—takes part in reproduction, the resulting child will have the wrong number of chromosomes, and this may lead to serious defects.

The presence of a certain extra chromosome, for instance, leads to the affliction called mongoloid idiocy, characterized not only by severe mental retardation but also by typical visible changes: the face is round and flat with oblique eyes, small mouth, and small, flattened ears. Lack of part of another chromosome also leads to imbecility, facial peculiarities, and an abnormality in the larynx giving these children a mewing voice (whence the name *cri du chat* syndrome).

Some particularly remarkable changes arise from

abnormalities in the number of sex chromosomes, which, as may be remembered, are XX in women and XY in men. For instance, the presence of an additional X chromosome in a man, who thus carries the set XXY, results in a marked "feminization." The testicles do not develop properly, no sperm is formed, and body and face possess feminine traits. This shows that the effect of the Y chromosome producing maleness is set off partly by that of the two X chromosomes.

The converse condition, an extra Y chromosome, might then be expected to produce a superman with the set XYY. Such a man would in fact be unusually tall, and suffer from slight debility, with a quite extreme tendency towards aggressiveness, which tends to make this unfortunate "superman" quite asocial. This is a particularly revealing instance of the genetic basis of behavior.

But such genetic changes with large, easily observed effects are at one end of the scale. Just as with other traits, it seems reasonable to assume that genes concerned with behavior may have effects varying greatly in size—from major to almost imperceptible. Furthermore, several different genes may affect the same trait, as in the case of skin color in man. Again, as with genes for physical traits, one gene may have several seemingly unrelated effects—the same genetic change may affect both physical and psychical characteristics, as in the case of the XYY "superman."

Study of such behavioral genes is still in an early stage, but enough has been done—for instance, in studies of twins (evaluated with the help of some particularly formidable mathematical methods)—to show that they are indeed present, although functioning as personality determinants in general rather than influencing any specific actions.

DEATH
BY VIOLENCE

Many people today die a violent death, and all the evidence suggests that the number of such deaths will rise. This is probably true for all parts of the world, and the highly advanced countries are no exception.

Traffic accidents are now one of the main causes of death in technological countries, taking many more lives than wars, and the number of traffic deaths is increasing every year. As world population grows and traffic also grows, accidents will multiply at an even higher rate. This is easy to see, for with an increase in the number of people moving about, the number of possible encounters increases even more rapidly. For example, on an island with only three automobiles, three different car-to-car meetings are possible. With four cars, the number of possible meetings rises to six, with five cars to nine, with six to fifteen, and so on. With a sufficient number of vehicles on the move, the traffic accident might become the decisive factor in keeping the population increase at bay.

But traffic accidents do not strike indiscriminately. In the first place, they almost exclusively hit people who are on the move. People who stay at home are generally safe from traffic accidents, except for those very few who are mown down in their homes by wayward cars or crashing airplanes. The number of traffic accidents is in a direct ratio to the mileage covered, depending also,

156

of course, on the kind of vehicle used. So the effect becomes selective. Your genes may make you specially fit for a certain trade, and this trade may or may not require much traveling. The sedentary type is less exposed.

Once you are out on the highway or sidewalk, new selective factors are brought into play. What, for instance, is your basic attitude towards driving? Builders of automobiles give their products names that suggest dashing, aggressive driving. Swashbuckling may have been very useful to a medieval traveller, who would frighten off the ruffians and vagabonds he might encounter on the road, but in modern traffic it is more or less lethal. There can be little doubt that selection is on its way to fashioning future generations with more innate prudence than the present one and that this selection undoubtedly will affect the way we conduct ourselves on modern highways.

Pedestrians also fall victim to traffic accidents, and children are the most exposed group. Here again, the mortality is probably selective to some degree. There is some evidence that precocious children, whose inquisitiveness and brightness race ahead of their motor responses, tend to be especially vulnerable.

But traffic accidents tend not only to promote a prudent, sedentary generation. The sense organs and motor responses are also constantly being tested. Good eyesight and rapid reactions may mean the difference between life and death. In this respect, selection may also bring about, again through a more innate prudence, more care when using alcohol and drugs while driving. The old law of the jungle has been revived by the automobile.

Of course, the statistical likelihood of your being

hurt will vary according to different modes of transportation, and the selective effect is also variable. The most dangerous vehicle is the motor bike, with the motorcar a good second. In contrast, accidents in public transportation vehicles are less selective in effect.

You have braved the dangers and arrived at your destination—your job. Here you are beset by new hazards, and, again, their nature is dependent on the kind of job you have. A longshoreman, for instance, lives a comparatively dangerous life. According to a recent study, in dock work there are about a hundred accidents for each million work-hours, as compared to about thirty for truck drivers, twenty for engineers, ten for bus drivers, six for taxi drivers, and almost zero for newspaper boys. And the accidents do not happen quite indiscriminately. You stick your neck out once too often, and you've had it. An extra bottle of booze yesterday? A tendency to act before thinking? A quarrel at home this morning? All of these things can have an affect on our susceptibility to accidents.

At home, in fact, accidents are also particularly common. To the housewife, household accidents are *the* occupational hazards, and again it is carelessness that tends to be punished in the long run. While it is true that many household appliances are unsafe because of poor design, it is nevertheless also true that a large number of household accidents occur from carelessness while washing windows, foolish handling of electrical appliances or gas ranges, and so on. Being a child also has certain occupational hazards that often result in accidents in the home. For example, an adult might forget to place medicine out of the reach of inquisitive hands or a kettle with steaming water might be within the grasp of a small child, thus resulting in death or lifelong disablement. At home you relax, become careless, and fall down the stairs.

You tire of it all and decide to take the easiest way out—a jump, a bullet, too many sleeping pills. The rate of suicide in Western countries fluctuates around 0.1 and 0.2 per thousand but shows important variations, suggesting that its influence, although small, is highly selective. For example, the rate is fairly low among white male New Yorkers in the reproductive ages, but later rises rather steeply to values of up to 0.4 or 0.5 in older age groups. Negroes in New York, on the other hand, have a suicide rate that reaches a terrifying peak among young and early middle-aged men, reflecting the bitterness and frustration induced by their lot.

Then there is homicide. Nobody is predestined to become a murder victim, any more than a suicide, and yet there may well be sufficient genetic correlation to give this, too, a selective influence. The arrogant? The imprudent? The defenseless? The lonely? Though suicide or homicide may come to only one in a thousand, this is still a ripple in the gene pool, a wind that may bring change in the long run.

Moving further into the areas of the underprivileged, we meet new forms of crippling or death by violence. Now it is deprivation that becomes the main factor, and foremost, it is hunger. It was mentioned earlier that the composition of the world population as a whole is changing as a result of selection by differential fertility: birthrates in certain parts of the world are much higher than in others, and this increases their part of the common gene pool. But a higher birthrate may be offset by a rate of mortality even higher, especially among infants, and if current trends continue, it will probably occur more frequently.

When these populations increase, hunger tends to race ahead, for, at the same time, natural resources are being drained by predatory exploitation in various guises. Forests are felled for timber, or the natural vegetation is burned to make room for crops that rapidly exhaust the soil. The wild game is killed off to create more space for cattle that overgraze the pasture. Then the soil erodes, and fertile tracts change into wasteland.

If hunger becomes the main factor in keeping population down, it may be that only the very smart and selfish will survive, thus giving their genes as a legacy to future generations.

Finally, there is war. The big bomb kills everybody, and defoliation and burned-earth tactics are also more or less indiscriminately destructive. But in the line of fire, casual ties still tend to be the members of the armed forces themselves, and they consist largely of selected people: selected for physical fitness, educability, and (in some, but far from all cases) devotion. As a large part of the world is now in a state of more or less continuous war, this sort of selection is at work.

CHEMICALS
AND DRUGS

Physical accidents or violence may cripple or kill. Poison, intentionally or unintentionally administered, does the same thing. The more we surround ourselves with poisonous substances, the more we are likely to get one or more of them into our system. As soon as such substances are permitted to act on us, selection comes into play.

The ways in which poisons can affect us are best studied on small organisms with short generations. The reaction of such organisms to toxic material has been investigated intensively. Many diseases, for instance, used to be cured easily by antibiotic medications, but then some of the microorganisms concerned evolved resistant strains that could not be killed by the drugs at hand. New drugs had to be developed, but then they encourage new, resistant strains, and so the cycle continues, in a never-ending battle.

The great citrus orchards in California are attacked by various scale insects, which have long been combated by fumigation with hydrocyanic gas, known as an extremely strong poison. As early as 1914, however, strains of scale insects appeared that were not destroyed completely by the cyanide, and they have gradually spread to many orchards.

It was soon shown that the resistance was due to a single mutation. It would seem that this genetic change gave the resistant scale insects the ability to keep their spiracles (through which they breathe) closed for as long as thirty minutes, a time long enough to let the poison evaporate. Normal scales, in contrast, can hold their breath for only about a minute, and this is not long enough to counteract the cyanide. The resistant scales, then, do not have actual physiological immunity to the poison as such: they have merely developed the ability to hold their breath.

Now what happens when a new poison is introduced into human society? Let us think of a poisonous substance that man has been familiar with for untold generations—alcohol. It comes in various guises—wine, beer, mead, cider, pulque, sake, whisky, and so on—and has been used by human beings for thousands of years in most parts of the world. It must be assumed that this led to selection which, in various ways, tended to make men more or less resistant to the poisonous effects of the drug. For instance, a large and regular intake of alcohol will in due time affect the liver. In countries where wine is a daily drink, livers are probably, on the whole, more efficient in dealing with alcohol than are those in countries where the use is intermittent—for instance, in the form of a weekly binge. The same should hold for other physical effects of alcohol.

There is also a behavioral aspect. Part of the resistance is probably similar to the ability of the scale insect to hold its breath. The corresponding ability in man would be the ability to take it or leave it, i.e., to avoid becoming an alcoholic. And we know quite well that this kind of selection is still going on.

Who becomes an alcoholic? Anybody might, under sufficient stress, but the threshold is surely different for people with different genetic backgrounds, and this means selection. People who become alcoholics have reduced fertility, length of life, and ability to take care of their offspring. But in most parts of the world this kind of natural selection has been at work for countless generations, and it has led to a large measure of immunity. In those societies where alcohol has long been in use, the majority of people do not become alcoholics. Alcohol, it is said, is "integrated" into society. There is good reason to believe that this integration also has a genetical component.

We know what happens when alcohol is first introduced to people who did not know its use previously. South American Indians had alcohol; North American Indians did not. The effect on the latter was catastrophic.

Much of the extinction of North American Indians may well be due to their exposure to alcohol even though persecution by the invaders was the main factor. The contrast with South America, where the aboriginal type continues to predominate, is in any case striking.

Resistance to alcohol need not imply resistance to other drugs. Tobacco is a good example. This habit was introduced by American Indians at a relatively late date, and so is comparatively new to the European population and to non-Indian Americans. To see the result, we have only to look around. There are compulsive smokers everywhere. Obviously our resistance to this habit is very weak.

What about natural selection? Yes, we have that too. Reports on the ill effects of smoking on the human body are multiplying. Death from lung cancer is rising rapidly and striking compulsive smokers. Those who avoid this hazard may fall victim to thrombosis or other diseases of the circulatory system. This is selection, and you can imagine a vindictive smile on its Gorgonian head. The Indians are getting even.

Will selection take effect? Compulsive smokers have shorter lives than non-smokers, but do they die early enough to affect their reproduction and thus the behavior of future generations? Will the weaker health of the smoker mean a serious setback to the opportunities of his children? The answer is probably yes, for selection is an extremely dependable mechanism. All of this would be part of the picture of efficient natural selection. Geneticists have concluded that even such a minute difference in reproduction as 1/1000 (or 0.001) will take effect in the long run. Thus, if a given number of smokers have 1,000 children, and an equal number of non-smokers have 1,001, this will suffice to influence evolution. I think we can be hopeful that in another two or three thousand years, Western man may be able to handle his tobacco in a sensible way.

But let us not be dazzled by this prospect. Before we have even had time to start dealing with selection for tobacco resistance properly, other drugs are being introduced by diligent experimentalists. Take marijuana, which (like hashish and related drugs) has long been used by peoples in Arab countries and in India, but does not appear particularly problematic in these areas. Its use is said to be integrated into these societies, and it should be clear by now that this also has a genetic aspect. There is little serious addiction; genotypes that would fall victim to this have presumably been selected against for hundreds of generations.

It remains to be seen what will happen when its use is introduced massively in a society with a different genetic make-up. That there will be a phase of selection is obvious, but how severe and how long? It will probably take more than one generation to tell. Perhaps the use of hashish will turn out to have high selective efficiency —that is, it will rapidly ruin the individuals who become addicts. In that case, if we are lucky, five hundred years might be sufficient to produce the same degree of general immunity that we now enjoy against alcohol. Interbreeding with people belonging to hashish-using cultures would presumably be helpful and might further shorten the necessary time.

Unfortunately, this is not all. There are too many horses pulling in different directions. New drugs of varying potency are being put on the market constantly, legally or illegally, and contribute their share to the selection picture. Highly selective mortality induced by drug-taking may turn out to be another important factor acting to stop the population explosion. Crowding and stress are mounting; Nirvana beckons.

Through such intensive selection, perhaps some thousand years hence we will breed a kind of superman who will be able to take it or leave it: wine, LSD, aspirin, hashish—perhaps even sex. He will eat fire and chew glass. Do we think he is worth it?

LOOKING FORWARD

So much for the past and present: what about the future? In this geological instant, the future that lies immediately before us is dominated by the fateful increase in human population. It will certainly stop eventually, but how? Wholesale destruction by wars, leaving only a few stragglers with seriously damaged genetic make-ups? Selection among these might breed a new race of better men, and mankind might ultimately benefit; still, the experiment does not seem worthwhile. It has been thought that such a catastrophe could throw man back to the ape stage, but this can be discounted, for we were never apes, and, furthermore, there is no going back; evolution does not reverse its course.

Otherwise, the population increase might also be stopped by a sharp rise in mortality due to such factors as increasing accidents, pollution, drug addiction, and so on (perhaps with a number of minor wars thrown in). Or will the well-fed and powerful simply turn on their weaker brothers and kill them like vermin, as at Auschwitz and My Lai?

There are all sorts of possibilities, and most of them

look so gruesome that we hardly dare think of them. Perhaps the most abhorrent is when men, so to speak, cease to look upon each other as members of the same species, regarding others instead as apes, gooks, pigs, or what have you. Political satire is replete with representations of the enemy in animal guise—as a crocodile, a spider, a vulture, and so forth. Besides being unfair to the animals in question (which are really wonderful and fascinating), this breeds the sort of alien-species reaction that is conducive to such outrages.

Such cataclysmic possibilities darken the perspective of our immediate future. It may be that mankind has run its course and, in going, may exterminate life on this globe. Still, there must be other ways, and indeed we already have the scientific know-how to solve the problems. The question is a political one of how to work it out in practice, without resorting to the once-and-for-all type of final solution: the *endgültige Lösung*.

Perhaps the solution will be a compromise, with a blend of the humane and the bestial: a painful, blundering, yet somehow modestly successful compromise, in which enough love and compassion are saved to carry us through. This is probably the best that can be hoped for.

If we try to look beyond these scenes of impending disaster, assuming that mankind will survive them and go on living in civilized societies, we see a potential future many times longer than the past history of men and hominids that has been the topic through most of this book. The earth has now existed for about five billion years, and there appears to be no reason to assume that it could not continue to exist for at least as long. The sun is evidently among the stable stars in the galaxy of the Milky Way, and it will continue to keep the earth a habitable planet for a very long time to come.

It is up to us to keep our side of the bargain. This

means that we have to start planning, not just for our generation and the next, but with a potential future of many million years before our eyes. That this means complete recirculation of all raw materials—a complete end to pollution and predatory exploitation, and so on —is evident. But it means much more than that. It means we have to stop playing with fire. No doubt we can develop a bomb that will destroy the earth, and threaten our enemy with it (there are some characters in history who were fully capable of that), but we had better not. The risk may be only one in a million that the threat will be carried out, and yet if the situation is repeated often enough, the statistically improbable event will become probable and finally almost certain. Somebody will push the button in the end.

As to the men who will populate that world of the future, many entertaining speculations have been made. On the one hand, these man of the future have been pictured as wizened pygmies with atrophied limbs but gigantic, overdeveloped brains. (The loss of the wisdom teeth and outer toes is not unlikely.) On the other hand, better nutrition—and probably also increased inter-breeding of different stocks with resulting hybrid vigor —has led to a spectacular increase in stature within recent years, and the trend may continue to some extent, but certainly not to produce giants in the real sense of the word. That is to say, not on earth.

What about other planets? In the future we may be able to colonize Mars and the moon permanently, and perhaps to reach planets in other solar systems. In time, such operations will almost inevitably lead to the evolution of completely different species of men on different planets. The environments (even if artificial in part) will be so different from those on earth that entirely new selection pressures will develop.

A Martian colonist, for instance, though living in an artificial atmosphere, would probably be under selective pressure to increase body size and cold tolerance. The low gravity of Mars makes it possible to attain a much larger size than on earth without undue stress on bone and muscle. And the paleontological record on earth shows that if size increase is possible, it will probably occur.

The reasons are simple enough. With increasing size, you reduce the surface area/body weight ratio, thus making you more resistant to cold. You also need less food and oxygen in relation to your body weight—in short, you live more thriftily. At the same time, you tend to live longer, probably because of a slower metabolic rate. A mouse crams as many heartbeats into its few months of existence as an elephant does into its seventy-odd years: you might say that both live just as much, but the mouse does it at a far more rapid rate.

This factor can also be seen in our own lives. Long ago, Gaston Backman pointed out that organismic time is logarithmic. In a logarithmic scale, the higher the numbers, the shorter the units; in the life of an organism, time passes faster with increasing age. This is a familiar truth to all of us. A day in childhood is a long, long time, filled with events and experiences ("Won't tomorrow ever come?"). In old age it comes only too soon. Where a childhood year floats blissfully on, eternity-like, the years pass ever more briskly as you grow older.

So the length of our life becomes, in a way, subjective. Perhaps the mouse could insist that its life really was as long as that of the elephant. And the hypothetical Martian of future millennia might feel that his life, though much longer in actual years, was really no longer than the earthman's; he only lived it at a slower rate.

With such adaptive changes taking place, we can see how crossing between earthmen and Martians would gradually come to an end. The traffic between the two planets could never become really intense—the trip would take months and would be excessively costly. The introduction of earth genes in a Martian colony would only produce misfits, and the same, of course, would be true for Martians coming to earth. (Possible hybrid vigor would be overshadowed by poor adaptation.) It is the arctic and tropic again, only much more so. Selection would now act to produce mating barriers between the two populations: they would become distinct species. The same would apply to the plants and animals that we might introduce.

As we have seen, this is nothing new in the evolutionary history of man. The stock has split up into different species before. But in the past all of them except one have ended by becoming extinct. The colonization of other planets, however, would give each species a niche of its own and thus a good chance of survival—unless, in our fine old tradition, we end up having great interplanetary wars.

Pursuing the theme still further, science fiction is fond of so-called humanoids—man-like creatures of alien origin, evolving separately on some distant planet and showing an amazing convergence with *Homo sapiens* (the males being sinister, the females beautiful). What is the actual likelihood that a man-like being would evolve on another planet? According to some recent cosmologies, solar systems like ours are not uncommon in the universe, and so there may be a large number of habitable planets in our region of the galaxy.

Man is a result of an evolution vastly longer than the part of it that has been traced in this book. Given an ancestral form like *Propliopithecus* and a suitable environment, the emergence of a man-like being may have had a rather high probability, though if things had gone differently, the robust Dartian might have been the surviving end form. However, in the total picture, almost everything had already been achieved with the emergence of *Propliopithecus*. Nature had by then produced (1) life, (2) unicellular organisms, (3) multicellular organisms, (4) animals with a notochord, (5) vertebrated animals, (6) land-living vertebrates, (7) mammals, (8) primates, and (9) higher primates of the basal hominid stock. At each step countless possibilities were present for further evolution, but only a small fraction of them were realized as surviving stocks of life—they had to be the right ones.

The emergence of man, then, is the result of an immensely long, complicated historical process, where sheer chance has played an important part in establishing some of our characteristics. Our teeth, for instance, began their history as the scale-like body covering of some obscure fish-like creatures in the seas of more than 400 million years ago. If they had not had these particular scales, the entire history of mammals would probably have been very different, if not impossible, as teeth have been of particular importance in that history.

Again, our arms and legs are derived from the pectoral and pelvic fins of the crossopterygian fishes that are the ancestors of all land-living vertebrates. If these fish had had another pair of fins, the land vertebrates might have been six-legged, like insects. Actually, this would make a semi-erect position with free hands much easier to attain, so that centaur-like beings might have evolved from more than one stock. Interplanetary tourists might be told to look out for such "humanoids."

A famous example of a characteristic that just happened to be there, but that was essential in our evolution, is the presence of two bones in the forearm—the radius and ulna. These bones form a very ingenious joint that makes it possible for us to rotate the hand around its long axis. This movement is necessary for almost any kind of manipulation. Without it, we would have no manual skill to speak of. So all of man's evolution really hinges upon the fact that there happened to be two bones—not just one—in this special part of the cross-opterygian fin. And yet, to that fish itself, these bones probably did not mean much. They were just part of a flexible fin that could also be used for propulsion on land.

Such historical accidents have played an important role in the shaping of our destiny, and it is most improbable that the total sequence could be re-enacted in any other case. If astronauts in a distant future are confronted by man-like "humanoids," they will probably turn out to be descended from human interplanetary colonists.

There is an exception, though. If the universe is infinite, there will be an infinite number of earths like ours, and an infinite number of copies of you, me, and all of us. But the infinite-universe cosmology has fared rather badly in recent years, and we may have to shelve this rather awesome idea. In another recently developed cosmology, the universe is formed by equal parts of matter and anti-matter. To make the symmetry perfect, we would have to believe in the presence of an anti-earth somewhere, populated by our doubles and consisting of anti-matter.

Generally speaking, however, intelligent beings can probably evolve in various different ways. If they live in a buoyant medium, they will probably be fish-like rather than man-like in external appearance. Recent studies of whales and dolphins make it quite clear that life on land, an upright posture, and skilled hands are not necessary prerequisites for high intelligence.

And here our attempt to look into the future and into distant worlds must come to an end. There is, however, one factor that has not been mentioned so far, but that may become decisive. Biological research is unraveling the genetic code and may in the future reach the point at which we can begin to control our evolution directly, instead of by the painful and roundabout way of artificial or natural selection. All populations in nature are on the horns of a dilemma: they need selection to keep healthy and viable, but selection is painful to the individual. If man can emancipate himself from that dilemma, he will have won real freedom—and this will be a unique situation in the history of life on earth.

In this book, I have tried to show that man's uniqueness does have a historical explanation. For more than 35 million years, our lineage has been a separate one. Julian Huxley thinks that it is unique in another sense, too: it is the only lineage which has a future evolutionary potential, the only one that can still give rise to something really new. A horse, a cat, a bat, etc. cannot evolve into anything really different; only men can do that. Perhaps he is right.

If so, let all of us partake. There are now three great races of man on earth—the black, the white, and the yellow. We rule the earth, but we are not alone; there are others: the Bushmen, the Australoids, and so forth. As A. Thoma believes, not a single leaf of the great tree of mankind should be lost, for there may be a role for all of us in the future.

HOMINIDS IN THE PLEISTOCENE ICE AGE

Glacial Chronology	Europe	Asia	Africa	Age (years)
Flandrian Interglacial	*Homo sapiens*	*Homo sapiens*	*Homo sapiens*	
Weichsel II Glaciation		*Homo sapiens* Skhul		10 000
Interstadial	*Homo sapiens*	Great Niah Cave	Broken Hill (*Homo erectus*)	30 000
Weichsel I Glaciation	Neanderthalers	Tabun, Solo	Omo, Kanjera (*Homo sapiens*)	40 000
Eemian Interglacial	Neanderthalers Fontéchevade (*H. sapiens*)	Mapa		70 000
Saalian Glaciation	Montmaurin			100 000
Holsteinian Interglacial	Swanscombe, Steinheim	Choukoutien		250 000
Elsterian Glaciation	Vertesszöllös		Ternifine, Kromdraai	
Cromerian Interglacial	Mauer		Olduvai (*Homo erectus*)	500 000
Günzian Glaciation		Lantian? Trinil	Swartkrans	
Villafranchian		Djetis		1 000 000

THE HIGHER PRIMATES IN THE GEOLOGICAL TIME-SCALE

Geological Epoch	Important Localities	Fossil Primates		Age Mill. Yrs.
		Ape-like	Man-like	
Pleistocene (Ice Age)	Olduvai	Modern Apes	Homo	
Pliocene	Omo	Gigantopithecus	Australopithecus (Dartians)	3
	Siwaliks	Dryopithecus, Pliopithecus	Ramapithecus	
Miocene	Ft. Ternan		Ramapithecus	12
	Rusinga	Dryopithecus, Pliopithecus		
		Aegyptopithecus		
Oligocene	Fayum	Oligopithecus	Propliopithecus	25
Eocene			Prosimians	35

179

GLOSSARY

(*Note: The definitions used in this glossary are based on the much more complete glossary in Frederick S. Hulse's book,* The Human Species: An Introduction to Physical Anthropology, *Second Edition (New York, 1971). I have, however, modified some of the definitions and added others to reflect my own views; therefore Professor Hulse should be warmly thanked for permitting use of his material, but absolved from responsibility for the version appearing here. I have also listed some commonly used Latin names for fossil hominids. For those regarded as synonyms, the valid name is given in parenthesis.*)

Abbevillian   The toolmaking tradition which preceded and evolved into the Acheulian (*q.v.*). The typical tools are coarsely made bifacial cutting implements of stone.

Acheulian   A toolmaking tradition of the Lower Paleolithic period. The typical tools are finely made bifacial cutting implements of stone.

adapt   To come to possess a genetic system suitable for existing ecological conditions. Populations adapt.

allele   One of two or more forms of a gene.

Anthropoidea   The suborder of primates containing apes, monkeys, and men.

ape   Common name of the family Pongidae.

Atlanthropus mauretanicus   Ternifine man (*Homo erectus*)

Australopithecus   A late Pliocene to middle Pleistocene hominid genus.

Australopithecus africanus   Gracile Dartians.

175

Australopithecus prometheus   Gracile Dartians from Makapan (*Australopithecus africanus*)

Australopithecus robustus   Robust Dartians.

brachiation   Moving about in the trees by swinging by the arms underneath branches.

breeding population   The group within which most mating takes place, and within which one is as likely to mate with any one person as with any other person of the opposite sex.

Cambrian   The first period of the Paleozoic era, between 500 and 400 million years ago.

Carboniferous   The fifth period of the Paleozoic era, from more than 200 million back to about 300 million years ago.

carbon-14 dating   A technique for dating the time at which an organism died by measuring the degree to which its carbon 14, which is radioactive, has vanished.

Cenozoic   The geological era which began about 65 million years ago, and in which we are living.

chromosome   A threadlike structure within the nucleus composed of or containing the genes. Except in gametes, chromosomes occur in pairs.

class   A taxon more inclusive than an order and less than a phylum. Members of a class share basic structural similarities although they may have undergone great adaptive radiation.

colugo   Also called "flying lemur," an insectivorous mammal with large skin folds for gliding.

convergent evolution   Evolution by which two groups become more similar to each other.

cranial capacity   The volume of the skull: this approximates the volume of the brain. Consequently, measuring it is the best means available for estimating how large the brains of extinct creatures, including our ancestors, may have been.

Cretaceous   The last period of the Mesozoic era, almost as long as all succeeding time.

Cromer   The interglacial between the Günz and Elster glaciations.

culture   Behavior which is learned, shared in society, and transmitted from one generation to the next.

Cyphanthropus rhodesiensis   Broken Hill man (late *Homo erectus*).

Dartian   Vernacular name for the genus *Australopithecus*.

dominance   In genetics, having an allele which is expressed in the phenotype even if heterozygous. In ethology, having a social position which gives one unquestioned priority in access to whatever is desired.

drift   In genetics a shift in allele frequencies which is due to chance alone—like a run of good or bad luck in gambling. Consequently, it is most likely to be effective in quite small populations.

Dryopithecus   A Miocene and Pliocene genus of apes which may be ancestral to the living large pongids.

ecology   The study of the mutual relations of organisms with one another and with the nonliving environment.

ecological niche   The position which a creature occupies in relation to its environment; its way of life as that fits into the living community of which it is a part.

Eemian   The interglacial between the Saale and Weichsel glaciations.

Eoanthropus dawsoni   Piltdown man, exposed as a hoax; no nomenclatural standing.

Elster   The second of the four Eurasian glaciations.

Eocene   The second epoch of the Cenozoic.

ethology   In zoology, the study of the naturalistic behavior of animals in groups.

evolution   Descent with modification. A shift in the allele frequencies of a population as time goes on.

family   A taxon more inclusive than a genus and less than an order. Members of a family share many structural peculiarities, as for example the Canidae, which includes dogs, wolves, and foxes.

femur, pl. femora   The long bone of the thigh.

fibula   The smaller of the two bones of the lower leg.

foramen, pl. foramina   A hole, specifically an opening in a bone.

foramen magnum   The opening in the skull through which the spinal cord emerges.

fossil   A remaining part of any ancient form of life. Commonly it is a hard part which has become mineralized, but prints or casts, when preserved, are also called fossils.

frontal bone   The bone of the forehead, extending back to form the foremost part of the skull vault.

gene   The unit of inheritance. At present it is commonly supposed that each gene is concerned with the synthesis of a particular enzyme.

genotype   The genetic constitution which reacts with the environment to produce the phenotype.

genus, pl. genera   A taxon more inclusive than a species and less than a family. The species of a genus share so many characteristics that they are likely to be in sharp competition.

gluteus maximus   In the Hominidae, the largest muscle of the hip and thigh; its strength permits erect posture without conscious strain. In other primates, it is smaller and weaker, and its course somewhat different.

Guanches   Aboriginal inhabitants of the Canary Islands, noted for their tall stature and extraordinary strength.

Günz   The first of the four Eurasian glaciations.

Holstein   The interglacial between the Elster and Saale glaciations.

Hominidae   The family including all species of *Homo* as well as the *Australopithecus* and *Ramapithecus*.

Hominoidea   The superfamily including the Hominidae and the Pongidae: that is, men and apes.

Homo   The genus to which we belong.

Homo erectus   The extinct species directly ancestral to *Homo sapiens.*

Homo habilis   Olduvai Bed 1 man, transition between *Australopithecus africanus* and *Homo erectus.* Early specimens probably belong to the former; the later ones to the latter.

Homo modjokertensis   Djetis man (*Homo erectus*)

Homo mousteriensis   Neanderthal man (*Homo neanderthalensis*)

Homo sapiens   Modern man. The only living species of the genus *Homo.*

Homo soloensis   Solo man (late *Homo erectus*).

Homo steinheimensis   Steinheim man (early *Homo sapiens* or *Homo neanderthalensis*).

humerus   The bone of the upper arm.

ilium, pl. ilia   The uppermost flaring bone of the pelvis which, during growth, fuses with the other pelvic bones.

ischium, pl. ischia   The lowermost bone of the pelvis which during growth fuses with the other pelvic bones.

Kenyapithecus wickeri   Fort Ternan *Ramapithecus* (*Ramapithecus punjabicus*).

mammal   The class of vertebrates possessing hair at some time during life, a lower jaw composed of a single bone, and, among females, means of suckling the young.

mandible   The lower jaw: in mammals a single bone. The mandible fossilizes much more readily than do many other bones.

Meganthropus africanus   East African gracile Dartian (*Australopithecus africanus*).

Meganthropus palaeojavanicus   Djetis robust Dartian (*Australopithecus robustus*).

Mesozoic   A geological era lasting from 200 million to 65 million years ago.

Miocene   The fourth epoch of the Cenozoic.

monkey   A grade of the Anthropoidea having tails. Two distinct taxa, the Ceboidea of the New World, and the Ceropithecoidea of the Old World, are of this grade.

moraine   Deposit formed in or by a glacier. End moraines form at the glacier's edge.

Mousterian   The major toolmaking tradition of the Middle Paleolithic, closely associated with Neanderthal man. A great variety of tools were made, including knives, scrapers, points, and augers.

mutation   A physical or chemical change in the structure of a gene which leads to a change in the gene's activity.

natural selection   Selection (q.v.) which takes place without intention or control by anyone, as contrasted to artificial selection by a breeder who wishes to produce a certain sort of creature.

Neanderthal   A valley in Germany where remains of a Pleistocene variety of man were found and after which this variety is named.

Neolithic   The period when men cultivated plants and made many of their tools of ground stone. Self-contained farming villages were typical.

notochord   A dorsally located stiffening rod, unsegmented, which precedes the backbone in embryonic development among vertebrates, but is more lasting in some other chordates.

Oldowan   A toolmaking tradition in Africa during Villafranchian times. The standard tool was a crudely made cleaver or chopper.

Oligocene   The third epoch of the Cenozoic.

order   A taxon more inclusive than a family but less than a class. Members of an order share adaptations to some broad ecological zone, but may differ as much as seals from cats, or people from lemurs.

Oreopithecus   An anthropoid genus of the Pliocene.

Osteodontokeratic   The alleged culture of the *Australopithecines*, who, Dart claims, made tools out of bones, teeth, and horn.

Paleoanthropus heidelbergensis Heidelberg (Mauer) man *(Homo erectus)*.

Paleocene The first epoch of the Cenozoic.

Paleolithic A geological era lasting from 500 million to 200 million years ago.

parallel evolution Evolution in the same direction by two related taxa, so that they do not become less similar despite lack of interbreeding, but instead both acquire similar new characteristics.

Paranthropus crassidens Kromdraai robust Dartian *(Australopithecus robustus)*.

Paranthropus robustus Swartkrans robust Dartian *(Australopithecus robustus)*.

Permian The final period of the Paleozoic era, about 200 million years ago.

phylogeny The evolutionary history of a taxon.

phylum, pl. phyla A taxon including all creatures which have a similar basic structural pattern, and which, therefore, are related to one another.

Pithecanthropus A fossil hominid genus of the lower middle Pleistocene. No longer used as a taxonomic title.

Pithecanthropus erectus Trinil man *(Homo erectus)*.

Pithecanthropus robustus Djetis man *(Homo erectus)*.

Pleistocene The sixth epoch of the Cenozoic, during which extensive glaciations took place in various regions. (Not synonymous with Paleolithic.)

Plesianthropus transvaalensis Sterkfontein gracile Dartian *(Australopithecus africanus)*.

Pliocene The fifth epoch of the Cenozoic.

pluvial A rainy period. Pluvial periods in tropical lands may have been contemporaneous with glacial periods in colder areas.

Pongidae The family including all the genera of apes.

potassium-argon dating A technique for dating the time at which a rock was formed by measuring the extent to which potassium has been transformed into argon.

preadaptation   A characteristic useful for ecological conditions in which a creature does not yet live.

primates   The order of mammals to which we belong.

Propliopithecus   An anthropoid genus of the Oligocene which may be ancestral to all hominids.

Propliopithecus haeckeli   Fayum hominid.

Prosimia   The suborder of primates containing lemurs, tarsiers, and other primitive forms.

race   A population within a species which can be readily distinguished from other populations on genetic grounds alone.

radius   One of the two bones of the forearm.

Ramapithecus   A late Miocene and early Pliocene hominid genus having teeth which strongly suggest close hominid affinities.

Ramapithecus punjabicus   Sole known species of *Ramapithecus*.

Saale   The third of the four Eurasian glaciations.

selection   The survival of some genotypes at the expense of contrasting ones, because the latter are less well adapted.

sex-linked   A gene is termed sex-linked if it is located on either the X or the Y chromosomes which determine sex. Its mode of inheritance is determined by the fact of its location.

sexual dimorphism   Marked difference in the characteristics of males and females of a species, for instance in body size.

sickle cell   A red blood cell which assumes a collapsed and distorted shape and is unable to carry the normal amount of oxygen. Such cells are due to an incompletely recessive allele: in heterozygous individuals relatively few sickle cells are found.

Sinanthropus pekinensis   Peking (Choukoutien) man (*Homo erectus*).

species, pl. species   The basic taxon among bisexual creatures. The lines of genetic communication are open
within a species, but only rarely *between* species. The
name of a species is always given in small letters after the
name of the genus to which it belongs.

taxon, pl. taxa   A group of creatures related by descent from
a common ancestor and distinctive enough to deserve a
name.

taxonomy   The science of systematic classification of living
things in such a manner as to indicate their relationship
to one another.

Tchadanthropus uxoris   Tchad man (late gracile Dartian or early
Homo erectus).

Telanthropus capensis   Advanced human from Swartkrans
(*Homo erectus*).

territorialism   In ethology, a concept often used to "explain"
the behavior of animals in the wild. It depends upon the
idea that defense of a creature's home territory is vital,
because it serves to space out individuals and thus con-
serve resources. Among primates, some species are
much more territorial than others.

thumb   A digit that is set apart from the others and facilitates
grasping. In most primates, thumbs are found on hands
and feet alike: a few lack thumbs on their hands, and we
lack thumbs on our feet.

tibia   The larger of the two bones of the lower leg.

ulna   One of the two bones of the forearm.

Villafranchian   The first part of the Pleistocene, before the
Günz glaciation. Modern species of animals begin to appear during this time.

Weichsel   The fourth and latest of the Eurasian glaciations.

Zinjanthropus boisei   Olduvai Bed 1 robust Dartian (*Australopithecus robustus*).

BJÖRN KURTÉN, a professor at the University of Helsinki, has worked extensively in Europe, Africa, and North America. He has published about eighty scientific articles, several volumes in English on mammals and the ice ages, and a number of novels in Swedish. He has held a lectureship at Harvard's Museum of Comparative Zoology and has been awarded a Rockefeller Foundation research grant, a University of Florida fellowship, and the Finnish State Prize for popular science writing.